W0082082

WAVE PROPAGATION
AND INVERSION

WAVE PROPAGATION AND INVERSION

Edited by W. E. Fitzgibbon
University of Houston

Mary Fanett Wheeler
Rice University

 Philadelphia

Society for Industrial and Applied Mathematics

WAVE PROPAGATION AND INVERSION

Library of Congress Cataloging-in-Publication Data

Wave propagation and inversion / edited by W. E. Fitzgibbon, Mary
  Fanett Wheeler.
         p.   cm.
     Includes bibliographical references (p.   ).
     ISBN 0-8971-300-5
     1. Wave-motion, Theory of.   2.   Inverse problems (Differential
equations)   3. Wave equations.   I. Fitzgibbon, W. E. (William
Edward), 1945–   .   II. Wheeler, Mary F. (Mary Fanett)   III. Society
for Industrial and Applied Mathematics.
QA927.W387   1992
530.1'24—dc20   92-16401

All rights reserved. Printed in the United States of America. No part of this book may
be reproduced, stored or transmitted without the written permission of the
Publisher. For information, write the Society for Industrial and Applied Mathematics,
3600 University City Science Center, Philadelphia, PA 19104-2688.

Copyright ©1992 by the Society for Industrial and Applied Mathematics.

PREFACE

In September of 1989, the Society for Industrial and Applied Mathematics, with the cooperation of the Department of Mathematics and the Energy Laboratory of the University of Houston, hosted a conference on mathematical and computational issues in geophysical fluid and solid mechanics. This was the third in an ongoing sequence of SIAM conferences pertaining to the geosciences.

The purpose of this conference was to provide a forum where mathematicians, geophysicists, geologists, hydrologists, and petroleum engineers could meet, discuss, and collaborate on problems of mutual interest. The central topics were systems of conservation laws, reactive flows, fluid and solid mechanics, partial differential equations, wave propagation, materials response, and geochemistry. Areas of application include flow in porous reservoirs and acquifers, basin modeling, seismic modeling and inversion contaminant transport, and remote sensing.

The program of the conference consisted of invited plenary lectures, minisymposia or invited special sessions, research workshop contributed papers, poster sessions, and informal seminars and discussions.

These volumes are not intended to be a proceedings of the conference per se. Participants in the conference were encouraged to submit manuscripts developed from the topics presented at the meeting. The submissions could either be expository papers or original research. The standard elongated abstract was not acceptable. All papers were refereed by outside reviewers.

The accepted papers fall roughly into three categories which we have somewhat arbitrarily labeled as Computational Methods in Geosciences, Modeling and Analysis of Diffusive and Advective Processes in Geosciences, and Wave Propagation and Inversion. These broad categories serve as the titles of this sequence of three volumes. We realize that our categorization is imperfect and hope that it does not serve to confuse or offend.

If dedications are appropriate, these volumes should be dedicated to Garrett T. Etgen, Chairman of the Mathematics Department at the University of Houston, who was tremendously supportive of the endeavor, worked endlessly, and derived no credit or visibility. In this dedication we show our deepest appreciation for his tireless efforts to further the development of mathematical sciences in the Houston area.

The errors that most assuredly occur in these volumes are consequences of the carelessness and incompetence of the editors and the editors herewith and henceforth apologize.

W. E. Fitzgibbon
Mary Fanett Wheeler

CONTENTS

CHAPTER 1

Waves in Partially Saturated Porous Media†

James G. Berryman*

Abstract. Efforts to extend the theory of poroelasticity to semilinear and non-linear elastic response, to partially saturated pores, to inhomogeneous solid frame materials, and to viscous losses due to localized flow effects are summarized. The prospects for a comprehensive theory of wave propagation in partially saturated porous media and the reasons for needing such a theory are also discussed. The main results are these: (*a*) Using the physically reasonable assumption of negligible capillary pressure change during passage of an acoustic signal through the medium, equations of poroelasticity for partially saturated materials have been derived and boundary conditions assuring the uniqueness of the solutions have been found. (*b*) Coefficients for scattering from a spherical inclusion in a poroelastic medium have been calculated. These coefficients may then be used to estimate effective constants in poroelastic wave equations when the medium is inhomogeneous; three common single-scattering approximations yield expressions that satisfy all known constraints on these constants and therefore provide generalized Gassmann's equations for inhomogeneous porous media. (*c*) The observed anomalously high attenuation of sound in partially saturated porous media can be explained in part by accounting for the effects of inhomogeneous porosity and fluid permeability. Regions of high permeability allow more fluid motion than regions of low permeability and therefore may be expected to play the dominant role in sound attenuation.

University of California, Lawrence Livermore National Laboratory, P. O. Box 808 L-202, Livermore, CA 94550.

†*Work performed under the auspices of the U. S. Department of Energy by LLNL under contract No. W-7405-ENG-48 and supported specifically by the Geosciences Research Program of the DOE Office of Energy Research within the Office of Basic Energy Sciences, Division of Engineering and Geosciences.*

1. Introduction. A limited theory of poroelasticity was formulated by Biot [1,2]. He assumed linear, isotropic elastic response on the macroscopic scale for porous media composed of homogeneous frame materials and fully fluid-saturated pores. The principal attenuation mechanism of this theory was viscous attenuation due to shear induced during macroscopic flow of the single-phase fluid filling the pores. Even with these simplifications, the resulting theory has remained a scientific oddity for over 30 years: (*a*) It is relatively hard to analyze the predictions of this theory [3–6] because it involves two coupled wave equations forming a system somewhat more complex than the equations of viscoelasticity [7,8] – which are nontrivial to analyze themselves! (*b*) The most startling predictions of the theory – such as the existence of a slow bulk compressional wave [1] or slow surface [9] and extensional waves [10] – are often very hard to verify in the laboratory [11–21]. (*c*) Even the validity of the form of the equations and the physical interpretation of many of the coefficients in the equations remained unclear for 25 years [12,22–27], and in some cases are still in dispute today [28,29]. It is therefore understandable that significant progress towards eliminating the many simplifying assumptions contained in the original work had not been made prior to the 1980s. Indeed, why complicate a subject which is already so difficult?

Often we try to argue that an elementary theory should suffice to explain the gross behavior of such complex materials, justifying our approximations with the comparative simplicity and elegance of the resulting analysis. If the theory is really successful at explaining the preponderance of experimental data, then of course our arguments are justified and it appears to be of only academic interest to expend such effort as would be required to construct a truly comprehensive theory. On the other hand, the theory to date has been unable to explain some of the most elementary experimental results for waves in geological materials, so it is essential to produce a more sophisticated theory capable of treating most of the complications encountered in practice. The need for a more realistic theory drives us to remove the simplifying assumptions. And this need for a more realistic theory becomes most apparent when we try to analyze sound wave data for real porous materials. For many of the geophysical applications of greatest interest, the pertinent geological materials are anisotropic and very heterogeneous, composed of multiple solid frame materials and multiple pore fluids. In some applications, the exciting waves are of large amplitude so that linear equations of motion are simply inadequate to describe the phenomena we want to study.

Various extensions of the elementary theory have been introduced. Biot himself had generalized the theory to include anisotropic effects for dynamic problems [30] and nonlinear effects for quasistatic problems [31]. When the saturating fluid is air [32–34], connections between Biot's theory and earlier work on rigid frame porous media [35] have also been explored. Various other authors have treated the generalization to partial saturation at very low frequencies in an intuitively appealing manner [36–40], but without having any clear procedure for generalizing their results for higher frequencies. The most general form of the equations for the elastic coefficients when the solid frame material is composed of two or more constituents has been known for some time [37], but no method for obtaining the required data had been suggested.

One goal of our research is a comprehensive theory of dynamic poroelasticity.

Irreversible pore collapse [41] is important in some of our applications, but we have neglected such effects initially in order to construct what is otherwise a quite general Lagrangian variational principle [42] for nonlinear and semilinear (reversible) deformations of dry and fluid saturated porous solids. This approach is very closely related to an Eulerian variational formulation of Drumheller and Bedford [43] for flow of complex mixtures of fluids and solids. We have shown that our theory reduces correctly to Biot's equations of poroelasticity [1] for small amplitude wave propagation and that it also reduces correctly to Biot's theory of nonlinear and semilinear rheology for porous solids [31] when the deformations are sufficiently slow. The resulting theory is a nontrivial generalization of Biot's ideas including explicit equations of motion for changes of solid and fluid density. Furthermore, if capillary pressure change may be neglected, the linear theory also shows that calculations on problems with only partially saturated pores may be reduced to computations of the same level of difficulty as those for fully saturated pores [44]. Appropriate boundary conditions have been found to guarantee that solutions of these equations are unique [44,45]. We expect the general theory to give a very good account of the behavior of wet porous materials during elastic deformations.

In the presentation that follows, we concentrate on three extensions of the theory of poroelasticity that tend to make the theory more realistic for applications to rocks. First, we show how the theory may be generalized to partially saturated porous media. Then, we use an effective medium method to find estimates of the coefficients in the equations when the frame material surrounding the pores is inhomogeneous. Finally, we analyze the attenuation of the fast compressional wave in heterogeneous media and show that the physically correct damping coefficient depends not on the global fluid-flow permeability, but on a spatial average (a line integral) of the local permeability.

2. Wave Equations for Multiple Fluid Saturation. When the mechanical and thermodynamical processes set in motion by a deformation are reversible, an energy functional which includes all the important effects involved in the motion may be constructed. Equations of motion may then be found by an application of Hamilton's principle. Such variational methods based on energy functionals are well-known in continuum mechanics [46]. Thus, the only really new feature in the present context is the degree of complexity; porous earth may be composed of many types of solid constituents and the pore space may be filled with a mixture of water and air. Some irreversible effects may also be included in the variational method (e.g., losses of energy due to drag between constituents) when they may be analyzed in terms of a dissipation functional. Other irreversible effects such as those associated with collapse of the pore space lie outside the scope of the traditional variational approaches; the forms normally used for the energy functionals are quadratic with constant coefficients in the linear problems or simply positive definite polynomials with constant coefficients for nonlinear problems. During pore collapse, the usual assumptions about the form of the energy functionals are violated, so the usefulness of the variational method is questionable. However, if we restrict discussion to linear or semi-linear processes, the variational methods are entirely adequate.

Using these variational methods, Berryman and Thigpen [44] have shown that the general equations of motion for linear elastic wave propagation through an isotropic porous medium containing both liquid and gas (or, more generally, any two fluids) in the pores are given by

$$
(1) \qquad \Lambda_{(\varepsilon)0}\ddot{\bar{\rho}}_{(\varepsilon)} = -\frac{\partial E_{(\varepsilon)}}{\partial \bar{\rho}_{(\varepsilon)}} - \frac{\lambda_\phi}{\bar{\rho}_{(\varepsilon)0}^2},
$$

$$
(2) \qquad
\begin{aligned}
&\rho_{(s)0}\ddot{u}_{(s)i} + \sum_{\gamma=g,l} \rho_{(s\gamma)0}\big(\ddot{u}_{(s)i} - \ddot{u}_{(\gamma)i}\big) \\
&\qquad = \Big[\rho_{(s)0}\frac{\partial E_{(s)}}{\partial u_{(s)i,j}} + \lambda_\phi \phi_{(s)0}\delta_{ij}\Big]_{,j} + d_{(s)i} + \rho_{(s)0}b_{(s)i},
\end{aligned}
$$

and

$$
(3) \qquad \rho_{(\gamma)0}\ddot{u}_{(\gamma)i} + \sum_{\xi\neq\gamma}\rho_{(\gamma\xi)0}\big(\ddot{u}_{(\gamma)i} - \ddot{u}_{(\xi)i}\big) = \phi_{(\gamma)0}\big(\lambda_\phi\big)_{,i} + d_{(\gamma)i} + \rho_{(\gamma)0}b_{(\gamma)i}
$$

where $\gamma = g$ or l and $\xi = g, l$, or s. The generalization to multiple pore fluids is immediate: let the index γ range over all fluids in the pores, and the index ξ range over all the fluids and the solid frame. The displacements are $u_{(\xi)i}$. The local densities (mass per unit volume of constituent) are $\bar{\rho}_{(\xi)}$. The partial densities (mass per unit total volume) are $\rho_{(\xi)} = \phi_{(\xi)}\bar{\rho}_{(\xi)}$. The internal energies of these immiscible constituents are $E_{(\xi)}$. The induced mass coefficients are $\rho_{(s\gamma)0}$. The body forces are given by $b_{(\xi)i}$ and the drag forces by $d_{(\xi)i}$. Thigpen and Berryman [47] have shown that the drag forces may be written in the form $d_{(\gamma)i} = -\sum_\xi D_{(\gamma\xi)}(\dot{u}_{(\xi)i} - \dot{u}_{(s)i})$ where $D_{(\gamma\xi)}$ is a symmetric, positive semidefinite matrix whose matrix elements satisfy $\sum_\gamma D_{(\gamma\xi)} = 0$ for $\xi = g$ or l. For the present discussion, we will ignore the effects of contact line motion that can be an added source of dissipation in partially saturated porous media [48,49]. We will also ignore possible effects of interaction torque that may lead to shear wave coupling between the pore fluids and the solid matrix [50].

One major simplification that occurs in the equations for partial saturation follows from (1) and the approximation $\Lambda_{(\xi)0} = 0$. We find that

$$
(4) \qquad \lambda_\phi = -\bar{\rho}_{(\xi)0}^2 \frac{\partial E_{(\xi)}}{\partial \bar{\rho}_{(\xi)}} \equiv -p_{(\xi)}
$$

where $p_{(\xi)}$ is the pressure for constituent ξ. Eq. (4) implies that all the pressures are equal – which is consistent with an assumption that capillary pressure effects are negligible for acoustics (also see References [51,52]). Without this approximation, the number of compressional waves through a porous medium will generally be one more than the number of fluids in its pores. This result is however dependent on the spatial arrangement of the fluids. If one fluid dominates and the others are mixed into the dominant one, then only two compressional waves are expected. When (4) is valid, only two compressional waves will be found regardless of the spatial arrangement of the fluids.

The subscript may subsequently be dropped from p. If $2e_{(\xi)ij} \equiv u_{(\xi)i,j} + u_{(\xi)j,i}$, then the first two strain invariants are defined by $I_{(\xi)1} = e_{(\xi)ii}$ and $I_{(\xi)2} = \frac{1}{2}[I_{(\xi)1}^2 - e_{(\xi)ij}e_{(\xi)ji}]$.

The changes in density are defined by $\Delta \bar{\rho}_{(\xi)} = \bar{\rho}_{(\xi)} - \bar{\rho}_{(\xi)0}$. In terms of these invariants, the standard definitions of the internal energies for an isotropic medium are

(5)
$$\rho_{(s)0} E_{(s)} = \frac{1}{2} a I_{(s)1}^2 + b I_{(s)2} + c I_{(s)1} \Delta \bar{\rho}_{(s)} + \frac{1}{2} d \Delta \bar{\rho}_{(s)}^2$$

and

(6)
$$\rho_{(\gamma)0} E_{(\gamma)} = \frac{1}{2} h_{(\gamma)} \Delta \bar{\rho}_{(\gamma)}^2$$

for $\gamma = g, l$. Applying (4) to (5) and (6), we find

(7)
$$p = \frac{\bar{\rho}_{(s)0}}{\phi_{(s)0}} (c I_{(s)1} + d \Delta \bar{\rho}_{(s)})$$
$$= \frac{\bar{\rho}_{(l)0}}{\phi_{(l)0}} h_{(l)} \Delta \bar{\rho}_{(l)}$$
$$= \frac{\bar{\rho}_{(g)0}}{\phi_{(g)0}} h_{(g)} \Delta \bar{\rho}_{(g)}.$$

The coefficients in (5) have been shown elsewhere [42] to be related to measurable quantities: $a = \phi_{(s)0} K^* / (\sigma - \phi_{(f)0}) + \frac{4}{3} \mu^*$, $b = -2\mu^*$, $c = \phi_{(s)0} K^* / \bar{\rho}_{(s)0} (\sigma - \phi_{(f)0})$, and $d = [\phi_{(s)0} / \bar{\rho}_{(s)0}]^2 K_{(s)} / (\sigma - \phi_{(f)0})$ where $\sigma = 1 - K^* / K_{(s)}$. The bulk and shear moduli of the drained porous solid frame are K^* and μ^*. The bulk modulus of the (assumed) single constituent composing the microscopically homogeneous frame is $K_{(s)}$. If the solid frame is composed of two or more constituents, then these formulas must be modified. The coefficient $h_{(\gamma)}$ is related to the bulk modulus $K_{(\gamma)}$ of the γ-th fluid constituent by

(8)
$$h_{(\gamma)} = \frac{\phi_{(\gamma)0}}{\bar{\rho}_{(\gamma)0}^2} K_{(\gamma)}.$$

These equations are all based in essential ways on Gassmann's equations [22, 37, 53]. Methods for generalizing these relations for isotropic porous materials will be presented in Section 5. The methods presented here could easily be generalized for anisotropic porous media, but at the present time little work has been done to identify appropriate measurements to determine the coefficients needed in the resulting equations so we will not pursue this line of research here.

Now we define the linearized increment of fluid content for partial saturation to be

(9)
$$\varsigma \equiv I_{(s)1} + \sum_{\xi} \frac{\phi_{(\xi)0}}{\bar{\rho}_{(\xi)0}} \Delta \bar{\rho}_{(\xi)}.$$

If only one fluid phase is present, (9) reduces to the exact result obtained previously [42]. If more than one fluid phase is present, then we observe that by defining an effective total fluid density change according to

(10)
$$\frac{\phi_{(f)0}}{\bar{\rho}_{(f)0}} \Delta \bar{\rho}_{(f)} \equiv \frac{\phi_{(g)0}}{\bar{\rho}_{(g)0}} \Delta \bar{\rho}_{(g)} + \frac{\phi_{(l)0}}{\bar{\rho}_{(l)0}} \Delta \bar{\rho}_{(l)}$$

with $\phi_{(f)0} \equiv \sum_\gamma \phi_{(\gamma)0}$ and we find that (9) reduces again to the exact result. Furthermore, applying (8), it is straight forward to show that (4) implies that

$$
(11) \qquad \frac{\Delta \bar{p}_{(\gamma)}}{\bar{p}_{(\gamma)0}} = \frac{p}{K_{(\gamma)}}
$$

for $\gamma = g$ or l. Substituting (11) into both sides of (10) shows that the effective bulk modulus of the multiphase fluid is given by

$$
(12) \qquad \frac{\phi_{(f)0}}{K_{(f)}} \equiv \sum_\gamma \frac{\phi_{(\gamma)0}}{K_{(\gamma)}}
$$

which is just the harmonic mean or Reuss average of the constituents' bulk moduli.

To check the consistency of our definition of ς, we can show easily that

$$
(13) \qquad \varsigma = \sum_\gamma \phi_{(\gamma)0}[I_{(\bullet)1} - I_{(\gamma)1}].
$$

If we define the average displacement of a fluid relative to the solid frame by

$$
(14) \qquad w_{(\gamma)i} = \phi_{(\gamma)0}[u_{(\gamma)i} - u_{(\bullet)i}]
$$

for $\gamma = g$ or l and the total relative fluid displacement by

$$
(15) \qquad w_i = \sum_\gamma w_{(\gamma)i},
$$

then (13) becomes

$$
(16) \qquad \varsigma = -w_{i,i}.
$$

Equation (16) reduces to the standard definition for full saturation when only one fluid saturates the pore space and is a natural generalization of this definition for partially saturated materials.

The total relative fluid displacement w_i defined by (15) is important in partial saturation problems not only because of the analogy just developed with the fully saturated problems, but also for convenience in applying boundary conditions in practical problems. Berryman and Thigpen [44] have shown previously that uniqueness of the solutions to the equations (1)-(3) demands the specification of either p or the normal component of this same w_i on the boundaries of the porous material. Therefore, it proves most convenient to combine these equations so that $u_{(\bullet)i}$ and w_i are the dependent variables. We will subsequently drop the subscript (s) on u_i since no confusion will arise and also define $e \equiv I_{(\bullet)1}$. In addition, the zero subscripts on density and volume fraction may also be dropped in the remainder of the analysis.

To determine the relations among $p, \varsigma,$ and e, substitute (11) and the first equation of (4) into (8) to eliminate $\Delta \bar{p}_{(\xi)}$ for all ξ. Using known identities and rearranging terms, we find easily that

$$
(17) \qquad p = M\varsigma - Ce
$$

where the coefficients C and M are given by

$$(18) \qquad C = \left\{ \left[(\sigma - \phi_{(g)} - \phi_{(l)})/K_{(s)} + \phi_{(g)}/K_{(g)} + \phi_{(l)}/K_{(l)} \right]/\sigma \right\}^{-1},$$

and

$$(19) \qquad M = C/\sigma$$

with

$$(20) \qquad \sigma = 1 - K^*/K_{(s)}.$$

Substituting (12) into (18) gives

$$(21) \qquad C = \left\{ \left[(\sigma - \phi_{(f)0})/K_{(s)} + \phi_{(f)0}/K_{(f)} \right]/\sigma \right\}^{-1}$$

which is the standard result for single-phase saturation [53].

Next we suppose the body forces vanish and sum the equations (2) and (3) to obtain

$$(22) \qquad \rho \ddot{u}_i + \bar{\rho}_{(g)} \ddot{w}_{(g)i} + \bar{\rho}_{(l)} \ddot{w}_{(l)i} = \left[\rho_{(s)} \frac{\partial E_{(s)}}{\partial u_{i,j}} - p \delta_{ij} \right]_{,j}$$

where $\rho = \sum_\xi \rho_{(\xi)}$. Dividing (3) through by $\phi_{(f)0}$ and rearranging terms, we find

$$(23) \qquad \bar{\rho}_{(g)} \ddot{u}_i + \frac{\alpha_{(g)} \bar{\rho}_{(g)}}{\phi_{(g)}} \ddot{w}_{(g)i} + \frac{D_{(gg)}}{\phi_{(g)}^2} \dot{w}_{(g)i} - \frac{\rho_{(gl)}}{\phi_{(g)}^2} \ddot{w}_{(l)i} = -p_{,i}$$

and

$$(24) \qquad \bar{\rho}_{(l)} \ddot{u}_i + \frac{\alpha_{(l)} \bar{\rho}_{(l)}}{\phi_{(l)}} \ddot{w}_{(l)i} + \frac{D_{(ll)}}{\phi_{(l)}^2} \dot{w}_{(l)i} - \frac{\rho_{(lg)}}{\phi_{(l)}^2} \ddot{w}_{(g)i} = -p_{,i}.$$

If an electrical tortuosity for a porous material is given by $\alpha = \phi F$ where ϕ is the effective porosity and F is the effective electrical formation factor (ratio of the conductivity of a conducting fluid to that of the insulating porous material when it contains the conducting fluid), then in (23) and (24) $\alpha_{(g)}$ is the electrical tortuosity of the pore space occupied only by the gas, while $\alpha_{(l)}$ is the electrical tortuosity of the pore space occupied only by the liquid. Introducing a Fourier time dependence of the form $exp(-i\omega t)$ into (23) and (24), combining, rearranging terms, and keeping the same names for the transformed and untransformed variables, we have

$$(25) \qquad -\omega^2 \begin{pmatrix} q_{(g)} & -r_{(g)} \\ -r_{(l)} & q_{(l)} \end{pmatrix} \begin{pmatrix} w_{(g)i} \\ w_{(l)i} \end{pmatrix} = \begin{pmatrix} -p_{,i} + \omega^2 \bar{\rho}_{(g)0} u_i \\ -p_{,i} + \omega^2 \bar{\rho}_{(l)0} u_i \end{pmatrix}$$

where

$$(26) \qquad \phi_{(\gamma)0}^2 q_{(\gamma)} \equiv \alpha_{(\gamma)} \rho_{(\gamma)0} + i D_{(\gamma\gamma)}/\omega$$

and

$$(27) \qquad \phi_{(\gamma)0}^2 r_{(\gamma)} \equiv \rho_{(gl)0}.$$

In (26), $\bar{\gamma} \neq \gamma$ so $\bar{\gamma} = l$ or g as $\gamma = g$ or l. Inverting the matrix in (25) and summing the results gives

(28)
$$-\omega^2[q_{(g)}q_{(l)} - r_{(g)}r_{(l)}]w_i = -(s_{(g)} + s_{(l)})p_{,i} + \omega^2[s_{(g)}\bar{\rho}_{(l)} + s_{(l)}\bar{\rho}_{(g)}]u_i$$

where

(29)
$$s_{(\gamma)} = q_{(\gamma)} + r_{(\gamma)}.$$

Using the expressions for $w_{(\gamma)i}$ from (25) again, we find

(30)
$$\sum_\gamma \bar{\rho}_{(\gamma)}w_{(\gamma)i} = \{[\bar{\rho}_{(g)}(q_{(l)} + r_{(g)}) + \bar{\rho}_{(l)}(q_{(g)} + r_{(l)})]p_{,i}$$
$$- \omega^2[q_{(l)}\bar{\rho}_{(g)}^2 + (r_{(g)} + r_{(l)})\bar{\rho}_{(g)}\bar{\rho}_{(l)}$$
$$+ q_{(g)}\bar{\rho}_{(l)}^2]u_i\}/\omega^2[q_{(g)}q_{(l)} - r_{(g)}r_{(l)}]$$
$$= \frac{\bar{\rho}_{(g)}(q_{(l)} + r_{(g)}) + \bar{\rho}_{(l)}(q_{(g)} + r_{(l)})}{s_{(g)} + s_{(l)}}w_i - \frac{(\bar{\rho}_{(g)} - \bar{\rho}_{(l)})^2}{s_{(g)} + s_{(l)}}u_i$$

where $p_{,i}$ has been eliminated in the second step of (30) using (28).

The final form of these equations is found by substituting (30) into (22), using (18) in the result and also in (28), and finally rearranging terms. The equations then take the familiar form [54,55]

(31)
$$\mu\nabla^2\vec{u} + (H - \mu)\nabla e - C\nabla\varsigma + \omega^2(\rho_{uu}\vec{u} + \rho_{uw}\vec{w}) = 0,$$

(32)
$$C\nabla e - M\nabla\varsigma + \omega^2(\rho_{wu}\vec{u} + \rho_{ww}\vec{w}) = 0,$$

where the inertial coefficients are given by

(33)
$$\rho_{uu} = \rho - \frac{(\bar{\rho}_{(g)} - \bar{\rho}_{(l)})^2}{s_{(g)} + s_{(l)}},$$

(34)
$$\rho_{wu} = \frac{\bar{\rho}_{(g)}s_{(l)} + \bar{\rho}_{(l)}s_{(g)}}{s_{(g)} + s_{(l)}} = \rho_{uw} + \frac{(r_{(l)} - r_{(g)})(\bar{\rho}_{(g)} - \bar{\rho}_{(l)})}{s_{(g)} + s_{(l)}},$$

and

(35)
$$\rho_{ww} = \frac{q_{(g)}q_{(l)} - r_{(g)}r_{(l)}}{s_{(g)} + s_{(l)}}.$$

The coefficient H is given by

(36)
$$H = K^* + \frac{4}{3}\mu^* + \sigma C$$

while C and M are given by (18) and (19). Thus, we find the remarkable result that the form of the equations of motion for partial saturation and for full saturation are the same – the only difference being that the inertial coefficients, as well as the C and M coefficients, are more complicated when the porous solid is only partially saturated.

3. Biot's Theory for Porous Materials with Inclusions. Now we will change notation somewhat and consider two isotropic porous media (i.e., host and inclusion) each of whose connected pore space is saturated with a single-phase viscous fluid. The fraction of the total volume occupied by the fluid is the void volume fraction or porosity ϕ, which is assumed to be uniform within a constituent but which may vary between the the host and inclusion. The bulk modulus and density of the fluid are K_f and ρ_f, respectively, in the host. The bulk and shear moduli of the drained porous frame for the host are K and μ. For now we assume the frame of the host is composed of a single constituent whose bulk and shear moduli and density are K_m, μ_m, and ρ_m. Corresponding parameters for the inclusion will be distinguished by adding a prime superscript. The frame moduli may be measured directly [55-57] or they may be estimated using one of the many methods developed to estimate elastic constants of composites [58,59].

For long-wavelength disturbances ($\lambda > h$, where h is a typical pore size) propagating through such a porous medium, we define average values of the (local) displacements in the solid and also in the saturating fluid. The average displacement vector for the solid frame is $\vec{u}$ while that for the pore fluids is $\vec{u}_f$. The average displacement of the fluid relative to the frame is $\vec{w} = \phi(\vec{u}_f - \vec{u})$. For small strains, the frame dilatation is

$$(37) \qquad e = e_x + e_y + e_z = \nabla \cdot \vec{u},$$

where e_x, e_y, e_z are the Cartesian strain components. Similarly, the average fluid dilatation is

$$(38) \qquad e_f = \nabla \cdot \vec{u}_f$$

(e_f also includes flow terms as well as dilatation) and the increment of fluid content is defined by

$$(39) \qquad \varsigma = \nabla \cdot \vec{w} = \phi(e - e_f).$$

With these definitions, Biot [1,2,30] shows that the strain-energy functional for an isotropic, linear medium is a quadratic function of the strain invariants [60] $I_1 = e, I_2$, and of ς having the form

$$(40) \qquad 2E = He^2 - 2Ce\varsigma + M\varsigma^2 - 4\mu I_2,$$

where

$$(41) \qquad I_2 = e_y e_z + e_z e_x + e_x e_y - \frac{1}{4}(\gamma_x^2 + \gamma_y^2 + \gamma_z^2),$$

and $\gamma_x, \gamma_y, \gamma_z$ are the shear strain components. Our earlier definitions (5) and (6) for partial saturation are completely consistent [42,61] with these definitions.

With time dependence of the form $exp(-i\omega t)$, the Fourier transformed version of the coupled wave equations of poroelasticity in the presence of dissipation take the form

$$(42) \qquad \mu \nabla^2 \vec{u} + (H - \mu)\nabla e - C\nabla\varsigma + \omega^2(\rho\vec{u} + \rho_f \vec{w}) = 0,$$

(43)
$$C\nabla e - M\nabla\varsigma + \omega^2(\rho_f\vec{u} + q\vec{w}) = 0,$$

where

(44)
$$\rho = \phi\rho_f + (1 - \phi)\rho_m$$

and

(45)
$$q = \rho_f[\alpha/\phi + iY(\xi)\eta/\kappa\omega].$$

The kinematic viscosity of the liquid is η, the permeability of the porous frame is κ, and the dynamic viscosity factor [2,62] is given (for our present choice of sign for the frequency dependence) by

(46)
$$Y(\xi) = \frac{1}{4}\xi T(\xi)/[1 + 2T(\xi)/i\xi],$$

where

(47)
$$T(\xi) = \frac{ber'(\xi) - ibei'(\xi)}{ber(\xi) - ibei(\xi)}$$

and

(48)
$$\xi = (\omega h^2/\eta)^{\frac{1}{2}}.$$

The functions $ber(\xi)$ and $bei(\xi)$ are the real and imaginary parts of the Kelvin function. The dynamic parameter h is a characteristic length generally associated with (and comparable in magnitude to) the steady-flow hydraulic radius. The electrical tortuosity α is a pure number related to the frame inertia which has been measured [27] for porous glass bead samples and has also been estimated theoretically [12,24]. The electrical tortuosity α and the fluid flow tortuosity τ are related by $\alpha = \tau^2 = \phi F$, where F is the electrical formation factor.

The coefficients H, C, and M are given by [37,53]

(49)
$$H = K + \frac{4}{3}\mu + \sigma C,$$

(50)
$$C = \{[(\sigma - \phi)/K_m + \phi/K_f]/\sigma\}^{-1},$$

(51)
$$M = C/\sigma,$$

where

(52)
$$\sigma = 1 - K/K_m.$$

The wave equations (42) and (43) decouple into Helmholtz equations for three modes of propagation if we note that the displacements $\vec{u}$ and $\vec{w}$ can be decomposed

(53)
$$\vec{u} = \nabla\Upsilon + \nabla\times\vec{\beta}, \quad \vec{w} = \nabla\psi + \nabla\times\vec{\chi},$$

where Υ, ψ are scalar potentials and $\vec{\beta}, \vec{\chi}$ are vector potentials. Substituting (53) into Biot's equations (42) and (43), we find they are satisfied if two pairs of equations hold:

$$(54) \qquad (\nabla^2 + k_s^2)\vec{\beta} = 0, \quad \vec{\chi} = -\Gamma_s \vec{\beta},$$

where $\Gamma_s = \rho_f / q$ and

$$(55) \qquad (\nabla^2 + k_{\pm}^2) A_{\pm} = 0.$$

In this notation, the subscripts $+, -$, and s refer respectively to the fast and slow compressional waves and the shear wave.

The wave vectors in (54) and (55) are defined by

$$(56) \qquad k_s^2 = \omega^2 (\rho - \rho_f \Gamma_s) \mu$$

and

$$(57) \qquad k_{\pm}^2 = (\omega^2 / 2\Delta)(b + f \mp [(b - f)^2 + 4cd]^{\frac{1}{2}}),$$

where

$$(58) \qquad b = \rho M - \rho_f C, \quad c = \rho_f M - qC, \quad d = \rho_f H - \rho C, \quad f = qH - \rho_f C,$$

with

$$(59) \qquad \Delta = MH - C^2.$$

The linear combination of scalar potentials has been chosen to be

$$(60) \qquad A_{\pm} = \Gamma_{\pm} \Upsilon + \psi,$$

where

$$(61) \qquad \Gamma_{\pm} = d / [(k_{\pm} \Delta / \omega^2)^2 - b] = [(k_{\pm} \Delta / \omega^2)^2 - f] / c.$$

With the identification (61), the decoupling is complete.

Since (54) and (55) are valid for any choice of coordinate system, they may be applied to boundary value problems with arbitrary symmetry. Biot's theory has therefore been applied to the scattering of elastic waves from a spherical inhomogeneity [4]. The results of that calculation will be summarized in the next section.

4. Scattering from a Poroelastic Spherical Inclusion.

The full analysis of scattering from a spherical inhomogeneity in a fluid-saturated isotropic porous medium is quite tedious. Fortunately, much of this work has already been done [4] and we may therefore merely quote the pertinent results here.

Let the spherical inhomogeneity (see Figure 1) have radius a. For the present, we place no restrictions on the properties of the inhomogeneous region. Thus frame bulk and shear moduli, grain bulk modulus, density, porosity, and permeability of a solid inclusion may all differ from those of the host. Furthermore, bulk modulus,

Figure 1. A spherical inclusion in a porous medium could be the result of local variations in fluid content, grain composition, porosity, permeability, etc.

density, and viscosity of the fluid in an inhomogeneous region may also all differ from those of the host fluid. Suppose now that a plane fast compressional wave is generated at a free surface far from the inclusion. Then, if the incident fast compressional wave has the form

$$\text{(62)} \qquad \vec{u} = \hat{z} \frac{A_0}{ik_+} \, exp\, i(k_+ z - \omega t),$$

the radial component of the scattered compressional wave contains both fast and slow parts in the far field and is given by

$$\text{(63)} \qquad u_{1r} = (ik_+)^{-1} exp\, i(k_+ r - \omega t)/k_+ r [B_0^{(+)} - B_1^{(+)} \cos\theta - B_2^{(+)}(3\cos 2\theta + 1)/4]$$
$$- (ik_-)^{-1} exp\, i(k_- r - \omega t)/k_- r [B_0^{(-)} - B_1^{(-)} \cos\theta - B_2^{(-)}(3\cos 2\theta + 1)/4].$$

Then, with the definitions $\kappa_\pm = k_\pm a$ and $\kappa_s = k_s a$ and with no restrictions on the materials, we find that

$$\text{(64)} \qquad B_0^{(-)} = \frac{i\kappa_-^3 A_0}{3M'(\Gamma_+ - \Gamma_-)(K' + \frac{4}{3}\mu)} \Big[(C - M\Gamma_-)(K' + \frac{4}{3}\mu)$$
$$- (C' - M'\Gamma_-)(K + \frac{4}{3}\mu) + (C - M\Gamma_-)(C' - M'\Gamma_-)\Big(\frac{C'}{M'} - \frac{C}{M}\Big) \Big],$$

and

$$\text{(65)} \qquad B_0^{(+)} = \frac{\kappa_+^3 A_0}{3i} \frac{[K' - K + (C - M\Gamma_-)(C'/M' - C/M)]}{K' + \frac{4}{3}\mu} + (\kappa_+/\kappa_-)^3 B_0^{(-)}.$$

Expansions of the other coefficients in the small parameter $\epsilon = C/K$ have been given in [4]. However, for the present application, only the first two coefficients are needed and these happen to be the ones known exactly at present. Of course, the full scattered wave also contains transverse components of the compressional wave, relative fluid/solid displacement, and mode converted shear waves. However, the scattering coefficients for these contributions are linearly dependent on the the the coefficients in (63) and therefore contain no new information. It is sufficient then to base our discussion on the expression (63).

As an elementary check on our analysis, we may first consider the limit in which the porosity ϕ vanishes. Then the fluid effects disappear from the equations and only the first line of (63) survives. Furthermore, it is not difficult to check [4] that the coefficients $B_n^{(+)}$ for $n = 0, 1, 2$ reduce to the well-known results for scattering from a spherical elastic inclusion in an infinite elastic medium [58]. For example,

$$(66) \qquad B_0^{(+)} = -i\kappa_+^3 A_0 (K' - K)/(3K' + 4\mu)$$

in this limit as expected.

These results have a multitude of potential uses. One straightforward application is the calculation of energy losses from elastic wave scattering by randomly distributed particles. A second important application is to use these results as the basis for an effective medium approximation for the effective constants of complex porous media. The second application is the one we address in the next section.

5. Generalization of Gassmann's Equations.

As noted previously, the equations of poroelasticity have several significant limitations. For example, these equations were derived with an explicit long-wavelength (low-frequency) assumption and also with strong implicit assumptions of homogeneity and isotropy on the macroscopic scale. Another restriction arises from the assumption that the pore fluid is uniform and that it fully saturates the pore space. For the present application, we assume that a single fluid saturates all the pore space for host as well as inclusion and scattering is caused by microscopic heterogeneity only in the solid properties.

Before deriving the main result, consider the problem of the porous frame without a saturating fluid (or with a highly compressible saturating gas). Then, since we take $C = M = \rho_f = 0$ in this limit, each term of Eq. (50) vanishes identically and the fluid dependent terms of Eq. (49) also vanish, leaving only the terms for the elastic behavior of the porous frame remaining. Since no slow wave can propagate under these circumstances, the second line of Eq. (63) disappears and only the fast wave terms contribute to the scattering. This limit is formally equivalent to the problem of elastic wave scattering from a spherical inclusion that has been treated in detail previously (see [58] and other references therein). One effective medium approximation (that we call the coherent potential approximation or CPA) requires the volume weighted average of the single-scattering results to vanish. This method simulates the physical requirement that the forward scattering should vanish at infinity if the impedance of the "effective medium" has been well matched to that of the composite. The resulting condition is that the volume weighted average of each of the $B_n^{(+)}$s for $n = 0 - 2$ must vanish. Using the convention that the effective constants for the composite porous medium are distinguished by an asterisk, the formulas for the effective bulk (K^*) and shear(μ^*) moduli for the drained porous frame of a microscopically heterogeneous medium are

(67)
$$\frac{1}{K^* + \frac{4}{3}\mu^*} = \left\langle \frac{1}{K(\bar{x}) + \frac{4}{3}\mu^*} \right\rangle$$

and

(68)
$$\frac{1}{\mu^* + F^*} = \left\langle \frac{1}{\mu(\bar{x}) + F^*} \right\rangle$$

where

(69)
$$F = (\mu/6)(9K + 8\mu)/(K + 2\mu).$$

The spatial($\bar{x}$) average is denoted by $\langle \cdot \rangle$. The remaining constant to be determined is the effective density which is just the average density [4]. Equation (67) follows easily from the volume average of (66), while (68) follows similarly from the volume average of B_2^+. Note that the equations for K^* and μ^* are coupled and therefore must be solved iteratively (i.e., self-consistently). Although the form of the equations (67) and (68) is identical to that obtained for elastic composites, the results can be quite different since the local constants $K(\bar{x})$ and $\mu(\bar{x})$ appearing in the formulas are frame moduli of the constituent spheres of drained porous material, not (necessarily) the moduli of the individual material grains. Of course, since the formula reduces correctly in the absence of porosity to the corresponding result for the purely elastic limit, the user of Eqs. (67) and (68) has some discretion about conceptually lumping grains together to form a porous frame or treating them as isolated elastic inclusions.

Now we will restrict discussion to the very low frequency limit where

(70)
$$\Gamma_+ = H/C$$

and

(71)
$$\Gamma_- = 0.$$

With these restrictions, the relevant scattering coefficients reduce to

(72)
$$B_0^{(-)} = \frac{i\kappa_-^3 C A_0}{3HM'(K' + \frac{4}{3}\mu)}\left[C(K' + \frac{4}{3}\mu + \sigma'C) - C'(K + \frac{4}{3}\mu + \sigma C)\right],$$

and

(73)
$$B_0^{(+)} = \frac{\kappa_+^3 A_0}{3i}\frac{[K' - K + (\sigma' - \sigma)C]}{K' + \frac{4}{3}\mu} + (\kappa_+/\kappa_-)^3 B_0^{(-)}.$$

The resulting conditions on the effective constants are

(74)
$$\left\langle \frac{\sigma^*[H(\bar{x}) + \frac{4}{3}\mu^* + \sigma(\bar{x})C(\bar{x})] - C(\bar{x})[K^* + \frac{4}{3}\mu^* + \sigma^*C^*]}{M(\bar{x})[K(\bar{x}) + \frac{4}{3}\mu^*]} \right\rangle = 0$$

and

(75)
$$\left\langle \frac{K(\vec{x}) - K^* + [\sigma(\vec{x}) - \sigma^*]C^*}{K(\vec{x}) + \frac{4}{3}\mu^*} \right\rangle = 0.$$

Recall that the averages in (74) and (75), as elsewhere in this paper, refer to spatial averages over (possibly) porous constituents of the overall porous aggregate. The limitations on the assumed geometry of the resulting aggregate have been discussed previously [63]. Note that (74) and (75) depend on the effective medium frame moduli K^* and μ^* determined by (67) and (68). The new constants determined by (74) and (75) are C^* and σ^*. The expressions for C^* and σ^* are coupled as written but may be decoupled after some algebra. The final expressions for these constants are

(76)
$$C^* = \sigma^* \Big/ \left[\left\langle \frac{1}{M(\vec{x})} \right\rangle + \left\langle \frac{\sigma^2(\vec{x}) - (\sigma^*)^2}{K(\vec{x}) + \frac{4}{3}\mu^*} \right\rangle \right]$$

and

(77)
$$\sigma^* = \left\langle \frac{\sigma(\vec{x})}{K(\vec{x}) + \frac{4}{3}\mu^*} \right\rangle \Big/ \left\langle \frac{1}{K(\vec{x}) + \frac{4}{3}\mu^*} \right\rangle.$$

Notice that both constants are determined explicitly by the formulas, in contrast to the frame moduli K^* and μ^* which are determined only implicitly by (67) and (68). The author has also shown [63,64] that (76) and (77) are completely consistent with all known constraints [37,53] on the form of these coefficients. Furthermore, the coherent potential approximation (CPA) treated here is just one of three approximations — including the average T-matrix approximation (ATA) and the differential effective medium (DEM) — all of which satisfy the known constraints on coefficients [64].

The same idea used to derive (76) and (77) has also been used to show [65] that the speed of waves propagating through a mixture of liquid and gas in the low frequency limit is given by Wood's formula [66] as expected [37,40].

Recently Berryman and Milton [67] have also derived exact results for generalized Gassmann's equations in composite porous media with two constituents. Some of these exact results were first obtained [64] using the single-scattering approach summarized here.

6. Anomalous Dissipation Caused by Inhomogeneous Fluid Permeability. A convincing demonstration has been given by Mochizuki [68] that, if we assume global fluid-flow effects dominate the viscous dissipation, Biot's theory of poroelasticity cannot explain the observed magnitude of wave attenuation in partially saturated rocks. Since the same theory explains the wave speeds quite well, it is reasonable to suppose that a small change in the theory may be adequate to repair this flaw. Many explanations are possible of course, but within the context of Biot's theory the simplest postulate is to suppose that local – rather than global – fluid-flow effects dominate the dissipation [28,29]. We will distinguish two related issues in this section which are summarized in the following questions: (a) Does the physics of wave propagation require that the value of the permeability κ appearing in Biot's equations should be that for global flow or for local flow? Then, if we can show that the value should be that for local flow, (*b*) does this change in the interpretation make enough difference so the theory can explain the correct magnitude for the attenuation?

To address the first question, we explore the consequences of assuming that Biot's theory should be applied at the local flow level rather than at the global flow level. This assessment is easily done by examining the dispersion relations. When the Fourier time dependence is $e^{-i\omega t}$ with angular frequency ω sufficiently low, Biot's theory predicts [see Eq. (57)] the dispersion relations for the fast (+) and slow (−) compressional modes in any homogeneous porous material to be

$$(78) \qquad k_+^2 \simeq \frac{\omega^2}{v_+^2}\left[1 + i\omega\frac{\rho_f^2}{\rho q_0}(1 - v_0^2/v_+^2)^2\right]$$

and

$$(79) \qquad k_-^2 \simeq \frac{i\omega q_0 H}{MH - C^2}$$

where

$$(80) \qquad v_+^2 = H/\rho,\, v_0^2 = C/\rho_f,\, q_0 = \rho_f\eta/\kappa.$$

The fraction of the total volume occupied by the fluid is the void volume fraction or porosity ϕ, which is assumed to be uniform. The bulk modulus and density of the fluid are K_f and ρ_f. The bulk and shear moduli of the drained porous frame are K and μ. For simplicity we assume the frame is composed of a single constituent whose bulk and shear moduli and density are K_m, μ_m, and ρ_m. Then the coefficients H, C, and M are given by (49)–(52). The overall density is

$$(81) \qquad \rho = \phi\rho_f + (1 - \phi)\rho_m.$$

The kinematic viscosity of the fluid is η and the permeability of the porous frame is κ.

Defining the quality factor for the fast compressional wave Q_+ by

$$(82) \qquad k_+^2 = \frac{\omega^2}{v_+^2}[1 + i/Q_+],$$

we find [69] that Q_+ is given by

$$(83) \qquad 1/Q_+ = \omega\frac{\kappa\rho_f}{\eta\rho}(1 - v_0^2/v_+^2)^2.$$

Since $1/Q_+$ is proportional to the permeability, the attenuation is therefore greatest in regions of high permeability. Thus, we might say that the regions of high permeability control the attenuation.

In the very low frequency limit, the slow compressional mode is known to reduce to Darcy flow with slowly changing magnitude and direction as the driving potential gradient oscillates sinusoidally [70–74]. Now consider a layered porous material (whose constants depend only on depth z) with constituents having identical physical constants except for the permeability κ which varies widely from layer to layer but which has a constant value κ_n within the n-th layer (lying in the range $z_{n-1} \leq z \leq z_n$ with $z_0 = 0$). Thus, the permeability is a piecewise constant function of z. The thickness of the n-th layer is given by $l_n = z_n - z_{n-1}$. If we impose a potential gradient along the z-direction in such a layered material, it is well-known that the effective permeability for fluid flow is found by taking the harmonic mean of the constituent permeabilities, $i.e.$,

$$(84) \qquad 1/\kappa_f = \frac{1}{L} \sum_n \frac{l_n}{\kappa_n},$$

where the total sample length L is given by the sum of the layer thicknesses

$$(85) \qquad L = \sum_n l_n.$$

From (84), we can conclude that the regions of lowest permeability dominate the effective overall permeability for fluid flow through a porous layered medium. Thus, we might say that the regions of low permeability $control$ the fluid flow – at least for this special choice of geometry.

The apparent attenuation of a fast compressional mode at normal incidence on such a structure has two distinct components: (a) Reflection and mode conversion at layer interfaces will have a tendency to degrade the fast wave, but this effect will be quite small at low frequencies for the model structure we are considering. (b) The attenuation within a layer is determined by the quality factor for that layer, as shown by Eq. (82). Assuming the attenuation is small enough, we may approximate (82) within any layer by $k_+(z) = (\omega/v_+)[1 + i/2Q_+(z)]$, where the functions $k_+(z)$ and $Q_+(z)$ take the piecewise constant values appropriate for the depth argument z. Neglecting the small effects of reflection and mode conversion, the behavior of the fast compressional wave at normal incidence is then easily seen to be given by

$$(86) \qquad A_+ exp\left[i \int_0^z dz k_+(z) - i\omega t\right] \simeq A_+ exp\left[i\frac{\omega}{v_+}z - i\omega t - \frac{\omega}{v_+}\int_0^z dz \frac{1}{2Q_+(z)}\right],$$

where A_+ is the amplitude of the wave at $z = 0$. In writing (86), we have used the piecewise constant property of the functions. The integral in the exponent is given by

$$(87) \qquad \int_0^z dz \frac{1}{2Q_+(z)} = \frac{\omega \rho_f}{2\eta\rho}(1 - v_0^2/v_+^2)^2 \int_0^z dz \kappa(z).$$

At the $z = L$ boundary of the material, we have

$$(88) \qquad \int_0^L dz \kappa(z) = \sum_n l_n \kappa_n.$$

If the layering is periodic with period much less than either z or L or if it is statistically homogeneous on this length scale, then we may approximate the integral in the exponent of (86) using (87) and

$$(89) \qquad \int_0^z dz\kappa(z) \simeq \kappa_a z,$$

where the effective permeability for attenuation measurements is given by the mean

$$(90) \qquad \kappa_a = \frac{1}{L}\sum_n l_n\kappa_n.$$

It is well-known that the mean is always greater than or equal to the harmonic mean of any function; thus,

$$(91) \qquad \kappa_f \leq \kappa_a.$$

In answer to our first query: the physics of wave propagation does dictate that local-flow effects dominate the attenuation of the fast compressional wave. The necessity of this conclusion is nicely illustrated in Figure 2. Suppose that a fast compressional wave is incident on a layered material with alternating permeable and impermeable layers. If the impermeable layers are very thin and have an acoustic impedance closely matching that of the permeable layers, their presence has a negligible effect on the propagating fast wave. The viscous attenuation of the fast wave occurs solely in the permeable layers and magnitude of that attenuation is completely determined by the permeability of these layers. By contrast, the global permeability of this material in the direction normal to the layering vanishes identically. If this null value were used in our predictions, the magnitude of the attenuation would be grossly underestimated. Although this choice of geometry is extreme, it clearly shows that errors in estimates of attenuation will arise if the value of permeability for global flow is used.

Now, can the theory predict the correct magnitude for the attenuation even with this change in the interpretation of the permeability factor? To predict the wave attenuation from measurements of permeability, we need some independent means of measuring the local permeability distribution. Normal laboratory flow experiments will not suffice, because they necessarily measure the global permeability. One promising method of estimating the local permeability uses image processing techniques to measure pertinent statistical properties of rock topology from pictures of cross sections [75,76]. This approach is still under development and we will not attempt to describe it in detail here.

Another approach, which is ultimately much less satisfactory than the image processing method but much easier to use at present, is to suppose that we can obtain reasonable estimates of the local permeability κ_L from the known values of the global permeability κ_G, the tortuosity $\tau = (\phi F)^{\frac{1}{2}}$, and the porosity ϕ. To do so requires some formula, so we will use a form of the Kozeny-Carman relation derived by Walsh and Brace [77]. For tubes of arbitrary ellipsoidal (major and minor axes a,b) cross-section the effective permeability of straight sections of such tubes is given by $\kappa = (\pi/4A)[a^3b^3/(a^2 + b^2)]$. The porosity for an ellipsoidal tube is

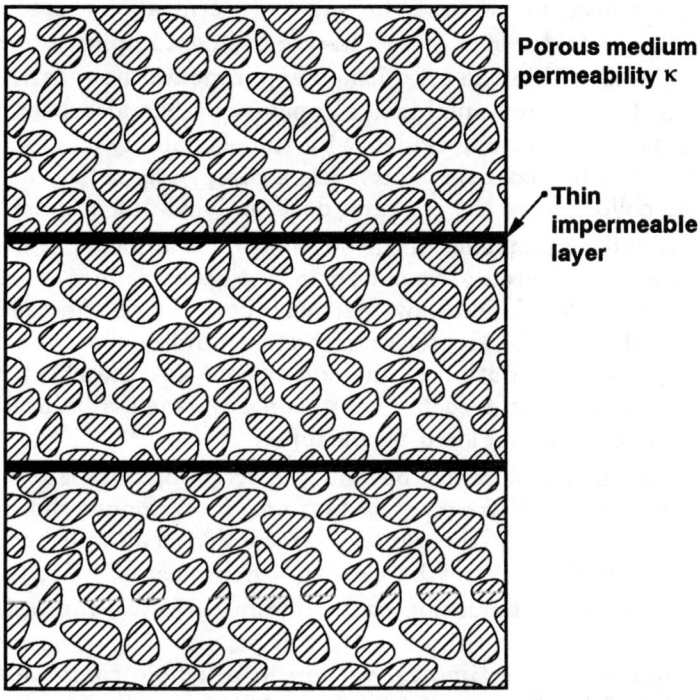

Figure 2. Illustration of a simple experiment to prove that the attenuation of the fast wave depends on the local – not the global – value of permeability κ. A fast wave incident normal to the impermeable partitions will experience a small but finite attenuation even though the global permeability in this direction vanishes identically.

$\phi = \pi ab/A$ and the specific surface area is well approximated by $s \simeq 2\pi[(a^2 + b^2)/2]^{\frac{1}{2}}/A$. Then, a Kozeny-Carman relation satisfied by κ, ϕ, and s can be shown to be

$$(92) \qquad \kappa = \frac{1}{2}\frac{\phi^3}{s^2}$$

for the effective permeability of a single tube oriented along the pressure gradient. If the tube is at an angle θ to this gradient, then Walsh and Brace [1984] show that

$$(93) \qquad \kappa = \frac{1}{2}\frac{\phi^3}{s^2\tau^2},$$

where $\tau = 1/cos\theta$. If we suppose that (92) and (93) are fairly representative of the material of interest, then (92) describes the maximum local permeability κ_L and (93) the effective global permeability κ_G. We then conclude that

$$(94) \qquad \kappa_L = \tau^2\kappa_G = \phi F\kappa_G.$$

The tortuosity τ has been measured for many sandstones; the values for samples studied by Simmons, Wilkens, Caruso, Wissler, and Miller [78–80] lie in the range

$2 \leq \tau \leq 5$, with most values $\tau \simeq 2$. To obtain estimates of attenuation close to experiment [39], we need to increase the value of permeability used in Mochizuki's calculations [68] by a factor of $\tau^2 \simeq 10$. This requirement implies a tortuosity of $\tau \simeq 3$, which is clearly well within the established experimental bounds. A more detailed analysis leading to the same qualitative conclusions has also been presented [69]. These arguments provide strong evidence for the plausibility of a local-flow explanation of the observed discrepancies. However, a completely satisfying demonstration must await the collection of the required data on local-flow permeability.

One unfortunate consequence of the observation that local permeability controls attenuation is that measured attenuation in wet rocks cannot be used directly as a diagnostic of the global fluid-flow permeability. Since the mean of the local permeabilities will always be greater than the true fluid-flow permeability regardless of the actual spatial distribution of the constituent κ's, the effective permeability computed from attenuation measurements can nevertheless be used to provide an upper bound on the desired global permeability.

Although the subject is really beyond the scope of this paper, in the context of the present volume on *Wave Propagation and Inversion* we should mention that indirect means of measuring the global permeability may still be viable. Seismic attenuation tomography [81] may be used to backproject effective local attenuation in a region from global attenuation data (line integrals). From estimates of local attenuation, we may deduce estimates of local permeability. Then, from a regional map of local permeability estimates, we can in principle compute the global permeability. Thus, although our results show that a simple direct measurement of global permeability is impossible, it certainly does not prevent us from obtaining the desired information from regional attenuation data.

7. Discussion. Why should we care about poroelasticity in general and slow waves in particular? Both the theory and the preponderance of experimental results have shown that, for earth materials containing some fluid at normal temperatures, the attenuation of slow compressional waves is so strong in the relevant frequency range (10–1000 Hz) that it is extremely unlikely that propagating slow waves will ever be directly observed in a field experiment. If we can ignore the slow waves, then the theory reduces to elasticity or viscoelasticity – which is clearly advantageous both conceptually and computationally. However, we miss something important if we try to compute wave propagation effects in the earth without using the equations of poroelasticity. The indisputable experimental evidence for the existence of the slow compressional mode in real materials [11–13,15,18,21,27,34] implies that mode conversions occur at every interface in a complex medium; a fast compressional wave striking any boundary (even at normal incidence) is partially reflected, partially transmitted, and partially converted into transmitted and reflected slow compressional modes [16]. Even if we never see a propagating slow wave in the field, the fast wave feels its presence as an additional attenuation mechanism that operates at every interface. The slow wave therefore gradually bleeds energy out of the propagating fast compressional wave into a highly damped viscous motion of fluid in the pores. Thus, the slow wave itself acts as an additional source of unaccounted for (and therefore anomalous) attenuation for those easily measured waves that do propagate. This attenuation mechanism is not predicted by the theories of elasticity or viscoelasticity. It may be possible to incorporate such effects into these simpler theories, but it seems more natural to use the theory that predicts the phenomenon. This is one practical reason why we should care about poroelasticity and why it is important to develop a comprehensive theory.

What then are the prospects for a comprehensive theory of poroelasticity? It appears likely that we will soon have a completely satisfactory linear theory of bulk waves including effects of partial saturation and inhomogeneous frame materials. A satisfactory nonlinear theory of bulk waves including effects of fracture, plastic flow, and pore collapse is at a more elementary stage, but is still likely to be achieved by the turn of the century. At present it appears that the most troublesome problems are those involving surface waves rather than the bulk waves. Surface waves depend critically on the nature of the equations of motion near interfaces. Using the standard boundary conditions of poroelasticity [44,45], it has been shown that a slow surface wave [9] or slow extensional wave [10] is expected only when a closed-pore boundary condition applies at the porous surface. Yet, available experimental data seem to show that such slow surface waves [18] do in fact propagate when the open-pore boundary condition applies. It is possible that the presence of a thin damage region close to the surface has a major effect on the conclusions of the theory regarding the propagation of the surface waves. However, it could also be that these experiments are pointing out still another subtle deficiency of the equations we use to describe wave propagation in porous media.

Acknowledgments. I thank S. C. Blair, B. P. Bonner, R. C. Y. Chin, G. W. Hedstrom, M. J. Miksis, G. W. Milton, and L. Thigpen for very helpful discussions and collaborations.

REFERENCES

1. M. A. Biot, *Theory of propagation of elastic waves in a fluid-saturated porous solid. I. Low-frequency range*, J. Acoust. Soc. Am., 28 (1956), pp. 168–178.

2. M. A. Biot, *Theory of propagation of elastic waves in a fluid-saturated porous solid. II. Higher frequency range*, J. Acoust. Soc. Am., 28 (1956), pp. 179–191.

3. R. Burridge and C. A. Vargas, *The fundamental solution in dynamic poroelasticity*, Geophys. J. Roy. Astr. Soc., 58 (1979), pp. 69–90.

4. J. G. Berryman, *Scattering by a spherical inhomogeneity in a fluid-saturated porous medium*, J. Math. Phys., 26 (1985), pp. 1408–1419.

5. A. N. Norris, *Radiation from a point source and scattering theory in a fluid saturated porous solid*, J. Acoust. Soc. Am., 77 (1985), pp. 2012–2023.

6. G. Bonnet, *Basic singular solutions for a poroelastic medium in the dynamic range*, J. Acoust. Soc. Am., 82 (1987), pp. 1758–1762.

7. R. C. Y. Chin, *Wave propagation in viscoelastic media*, in Physics of the Earth's Interior, Proceedings of the Enrico Fermi Summer School 1979, Course LXXVIII, North-Holland, Amsterdam, 1980, pp. 213–246.

8. T. Bourbié, O. Coussy, and B. Zinszner, *Acoustics of Porous Media*, Gulf, Houston, 1987, Chapt. 3.

9. S. Feng and D. L. Johnson, *High-frequency acoustic properties of a fluid/porous*

solid interface. I. New surface mode, J. Acoust. Soc. Am., 74 (1983), pp. 906–914.

10. J. G. Berryman, *Dispersion of extensional waves in fluid-saturated porous cylinders at ultrasonic frequencies*, J. Acoust. Soc. Am., 74 (1983), pp. 1805–1812.

11. T. J. Plona, *Observation of a second bulk compressional wave in a porous medium at ultrasonic frequencies*, Appl. Phys. Lett., 36 (1980), pp. 259–261.

12. J. G. Berryman, *Confirmation of Biot's theory*, Appl. Phys. Lett., 37 (1980), pp. 382–384.

13. D. Salin and W. Schön, *Acoustics of water saturated packed glass spheres*, J. Phys. Lett., 42 (1981), pp. 477-480.

14. R. Lakes, H. S. Yoon, and J. L. Katz, *Slow compressional wave propagation in wet human and bovine cortical bone*, Science, 220 (1983), pp. 513–515.

15. J. G. M. van der Grinten, M. E. H. van Dongen, and H. van der Kogel, *A shock-tube technique for studying pore-pressure propagation in a dry and water-saturated porous medium*, J. Appl. Phys., 58 (1985), pp. 2937–2942.

16. R. C. Y. Chin, J. G. Berryman, and G. W. Hedstrom, *Generalized ray expansion for pulse propagation and attenuation in fluid-saturated porous media*, Wave Motion, 7 (1985), pp. 43–66.

17. R. Lakes, H. S. Yoon, and J. L. Katz, *Ultrasonic wave propagation and attenuation in wet bone*, J. Biomed. Engng., 8 (1986), pp. 143–148.

18. M. J. Mayes, P. B. Nagy, L. Adler, B. P. Bonner, and R. Streit, *Excitation of surface waves of different modes at fluid-porous solid interface*, J. Acoust. Soc. Am., 79 (1986), pp. 249–252.

19. K.-J. Dunn, *Acoustic attenuation in fluid-saturated porous cylinders at low frequencies*, J. Acoust. Soc. Am., 79 (1986), pp. 1709–1721.

20. K.-J. Dunn, *Sample boundary effect in acoustic attenuation of fluid-saturated porous cylinders*, J. Acoust. Soc. Am., 81 (1987), pp. 1259–1266.

21. Q. Xue and L. Adler, *An improved method to measure slow compressional wave in fluid saturated porous plaates using Lamb modes*, in Review of Progress in Quantitative Nondestructive Evaluation, Vol. 9A, D. O. Thompson and D. E. Chimenti (eds.), Plenum Press, New York, 1990, pp. 211–218.

22. M. A. Biot and D. G. Willis, *The elastic coefficients of the theory of consolidation*, J. Appl. Mech., 24 (1957), pp. 594–601.

23. J. Geertsma and D. C. Smit, *Some aspects of elastic wave propagation in fluid-saturated porous solids*, Geophysics, 26 (1961), pp. 169–181.

24. R. J. S. Brown, *Connection between formation factor for electrical resistivity and fluid-solid coupling factor in Biot's equations for acoustic waves in fluid-filled porous media*, Geophysics, 45 (1980), pp. 1269–1275.

25. D. L. Johnson, *Equivalence between fourth sound in liquid He II at low temperatures and the Biot slow wave in consolidated porous media*, Appl. Phys. Lett., 37 (1980), pp. 1065–1067.

26. R. Burridge and J. B. Keller, *Poroelasticity equations derived from microstructure*, J. Acoust. Soc. Am., 70 (1981), pp. 1140–1146.

27. D. L. Johnson, T. J. Plona, C. Scala, F. Pasierb, and H. Kojima, *Tortuosity and acoustic slow waves*, Phys. Rev. Lett., 49 (1982), pp. 1840–1844.

28. J. G. Berryman, *Elastic wave attenuation in rocks containing fluids*, Appl. Phys. Lett., 49 (1986), pp. 552–554.

29. J. G. Berryman, *Seismic wave attenuation in fluid-saturated porous media*, Pure Appl. Geophys., 128 (1988), pp. 423–432.

30. M. A. Biot, *Generalized theory of acoustic propagation in porous dissipative media*, J. Acoust. Soc. Am., 34 (1962), pp. 1254–1264.

31. M. A. Biot, *Nonlinear and semilinear rheology of porous solids*, J. Geophys. Res., 78 (1973), pp. 4924–4937.

32. J. M. Sabatier, H. E. Bass, L. N. Bolen, K. Attenborough, and V. V. S. S. Sastry, *The interaction of airborne sound with the porous ground: The theoretical formulation*, J. Acoust. Soc. Am., 79 (1986), pp. 1345–1352.

33. K. Attenborough, *On the acoustic slow wave in air-filled granular media*, J. Acoust. Soc. Am., 81 (1987), pp. 93–102.

34. P. B. Nagy, L. Adler, and B. P. Bonner, *Slow wave propagation in air-filled porous materials and natural rocks*, Appl. Phys. Lett., 56 (1990), pp. 2504–2506.

35. C. Zwikker and C. W. Kosten, *Sound Absorbing Materials*, Elsevier, Amsterdam, 1949.

36. S. N. Domenico, *Effects of water saturation of sand reservoirs encased in shales*, Geophysics, 29 (1974), pp. 759–769.

37. R. J. S. Brown and J. Korringa, *On the dependence of the elastic properties of a porous rock on the compressibility of the pore fluid*, Geophysics, 40 (1975), pp. 608–616.

38. S. N. Domenico, *Elastic properties of unconsolidated sand reservoirs*, Geophysics, 41 (1977), pp. 882–894.

39. W. F. Murphy III, *Effects of partial water saturation on attenuation in Massilon sandstone and Vycor porous glass*, J. Acoust. Soc. Am., 71 (1982), pp. 1458–1468.

40. W. F. Murphy III, *Acoustic measures of partial gas saturation in tight sandstones*, J. Geophys. Res, 89 (1984), pp. 11549–11559.

41. J. F. Schatz, *Models of inelastic volume deformation of porous geologic materials*, in The Effects of Voids on Material Deformation, AMD – Vol. 16, edited by S. C. Cowin and M. M. Carroll, American Society of Mechanical Engineers, New York, 1976, pp. 141–170.

42. J. G. Berryman and L. Thigpen, *Nonlinear and semilinear dynamic poroelasticity with microstructure*, J. Mech. Phys. Solids, 33 (1985), pp. 97–116.

43. D. S. Drumheller and A. Bedford, *A thermomechanical theory for reacting immiscible mixtures*, Arch. Rational Mech. Anal. **73** (1980), pp. 257–284.

44. J. G. Berryman and L. Thigpen, *Linear dynamic poroelasticity with microstructure for partially saturated porous solids*, ASME J. Appl. Mech., 52 (1985), pp. 345–350.

45. H. Deresiewicz and R. Skalak, *On uniqueness in dynamic poroelasticity*, Bull.

Seismol. Soc. Am., 53 (1963), pp. 783–788.

46. A. Bedford, *Hamilton's Principle in Continuum Mechanics*, Research Notes in Mathematics, Vol. 139, Pitman, Boston, 1985.

47. L. Thigpen and J. G. Berryman, *Mechanics of porous elastic materials containing multiphase fluid*, Int. J. Eng. Sci., 23 (1985), pp. 1203–1214.

48. E. B. Dussan V., *Incorporating the influence of wettability into models of immiscible fluid displacement through porous media*, Physics and Chemistry of Porous Media II, AIP Conference Proceedings, Vol. 154, ed. J. R. Banavar, J. Koplik, and K. W. Winkler, AIP, New York, 1987, pp. 83–97.

49. M. J. Miksis, *Effects of contact movement on the dissipation of waves in partially saturated rocks*, J. Geophys. Res., 93 (1988), pp. 6624–6634.

50. Q.-R. Liu and N. Katsube, *The discovery of a second kind of rotational wave in a fluid-filled porous material*, J. Acoust. Soc. Am., 88 (1990), pp. 1045–1053.

51. J. E. Santos, J. M. Corberó, and J. Douglas, Jr., *Static and dynamic behavior of a porous solid saturated by a two-phase fluid*, J. Acoust. Soc. Am., 87 (1990), pp. 1428–1438.

52. J. E. Santos, J. Douglas, Jr., J. Corberó, and O. M Lovera, *A model for wave propagation in a porous medium saturated by a two-phase fluid*, J. Acoust. Soc. Am., 87 (1990), pp. 1439–1448.

53. F. Gassmann, *Über die elastizität poröser medien*, Vierteljahrsschrift der Naturforschenden Gesellschaft in Zürich, 96 (1951), pp. 1–23.

54. M. A. Biot, *Mechanics of deformation and acoustic propagation in porous media*, J. Appl. Phys., 33 (1962), pp. 1482–1498.

55. R. D. Stoll, *Sediment Acoustics*, Lecture Notes in Earth Sciences, Vol. 26, Springer, Berlin, 1989, Chapts. 4 and 5.

56. R. D. Stoll and G. M. Bryan, *Wave attenuation in saturated sediments*, J. Acoust. Soc. Am., 47 (1970), pp. 1440–1447.

57. R. D. Stoll, *Acoustic waves in ocean sediments*, Geophysics, 42 (1977), pp. 715–725.

58. J. G. Berryman, *Long-wavelength propagation in composite elastic media I. Spherical inclusions*, J. Acoust. Soc. Am., 68 (1980), pp. 1809–1819.

59. P. R. Ogushwitz, *Applicability of the Biot theory. I. Low-porosity materials; II. Suspensions; III. Wave speeds versus depth in marine sediments*, J. Acoust. Soc. Am., 77 (1985), pp. 429–464.

60. A. E. H. Love, A Treatise on the Mathematical Theory of Elasticity, Dover, New York, 1944, pp. 43, 62, 102.

61. A. Bedford and D. S. Drumheller, *A variational theory of porous media*, Int. J. Solids Structures, 15 (1979), pp. 967–980.

62. A. N. Norris, *On the viscodynamic operator in Biot's equations of poroelasticity*, J. Wave-Material Interaction, 1 (1986), pp. 365–380.

63. J. G. Berryman, *Effective medium approximation for elastic properties of porous solids with microscopic heterogeneity*, J. Appl. Phys., 59 (1986), pp. 1136–1140.

64. J. G. Berryman, *Single-scattering approximations for coefficients in Biot's equa-*

tions of poroelasticity, J. Acoust. Soc. Am., (1992), in press.

65. J. G. Berryman and L. Thigpen, *Effective constants for wave propagation through partially saturated porous media*, Appl. Phys. Lett., 46 (1985), pp. 722–724.

66. A. W. Wood, *A Textbook of Sound*, Bell, London, 1957, p. 360.

67. J. G. Berryman and G. W. Milton, *Exact results for generalized Gassmann's equations in composite porous media with two constituents*, Geophysics, 56 (1991), in press.

68. S. Mochizuki, *Attenuation in partially saturated rocks*, J. Geophys. Res., 87 (1982), pp. 8598–8604.

69. J. G. Berryman, L. Thigpen, and R. C. Y. Chin, *Bulk elastic wave propagation in partially saturated porous solids*, J. Acoust. Soc. Am., 84 (1988), pp. 360–373.

70. D. L. Johnson, J. Koplik, and R. Dashen, *Theory of dynamic permeability and tortuosity in fluid-saturated porous media*, J. Fluid Mech., 176 (1987), pp. 379–402.

71. P. Sheng and M.-Y. Zhou, *Dynamic permeability in porous media*, Phys. Rev. Lett., 61 (1988), pp. 1591–1594.

72. E. Charlaix, A. P. Kushnick, and J. P. Stokes, *Experimental study of dynamic permeability in porous media*, Phys. Rev. Lett., 61 (1988), pp. 1595–1598.

73. D. L. Johnson, *Scaling function for dynamic permeability in porous media*, Phys. Rev. Lett., 63 (1989), p. 580.

74. M.-Y. Zhou and P. Sheng, *First-principles calculations of dynamic permeability in porous media*, Phys. Rev. B, 39 (1989), pp. 12027–12039.

75. J. G. Berryman and S. C. Blair, *Use of digital image analysis to estimate fluid permeability of porous materials: Application of two-point correlation functions*, J. Appl. Phys., 60 (1986), pp. 1930–1938.

76. J. G. Berryman and S. C. Blair, *Kozeny-Carman relations and image processing methods for estimating Darcy's constant*, J. Appl. Phys., 62 (1987), pp. 2221–2228.

77. J. B. Walsh and W. F. Brace, *The effect of pressure on porosity and the transport properties of rock*, J. Geophys. Res., 89 (1984), pp. 9425–9431.

78. G. Simmons, R. Wilkens, L. Caruso, T. Wissler, and F. Miller, *Physical properties and microstructures of a set of sandstones*, Annual Report to the Schlumberger-Doll Research Center, 1 January 1982, pp. VI–16.

79. G. Simmons, R. Wilkens, L. Caruso, T. Wissler, and F. Miller, *Physical properties and microstructures of a set of sandstones*, Annual Report to the Schlumberger-Doll Research Center, 1 January 1983, pp. VI–16.

80. L. Caruso, G. Simmons, and R. Wilkens, *The physical properties of a set of sandstones — Part I. The samples*, Int. J. Rock Mech. Min. Sci. Geomech. Abst., 22 (1985), pp. 381–392.

81. J. R. Evans and J. J. Zucca, *Active high-resolution seismic tomography of compressional wave veolcity and attenuation structure at Medicine Lake Volcano, Northern California Cascade Range*, J. Geophys. Res., 93 (1988), pp. 15016–15036.

CHAPTER 2

A New Approach to Inverse Problems of Wave Equations (II) (The MCF Method in Time-Domain)

Hua Ding*
Shouze Xu**
Zhemin Zheng*

Abstract. We developed the MCF method, introduced in our previous paper, in the time-domain. And we have got better results. Comparing with other inversion methods the MCF method has many advantages: simple physical interpretation, good convergence, good stability, and be easily programed for multi-dimensional problems.

INTRODUCTION

The MCF (Multi-Cost-Functional) method introduced by the authors(cf.[1]) is a powerful method for solving inverse problems of wave equations. But since we worked in frequency-domain in [1] the full advantage of the MCF method could not really be developed. In other words, because of the exponential factor of the Laplace transform we could not separate properly the data for different times and the influence of earlier data always persists and is great. This can be overcome in the time-domain. Regarding at the numerical experiments that we have done we see that the MCF method did give us very satisfactory results.

The method we present here is general for inverse problem of different governing wave equations (examples: acoustic wave, elastic wave, visco-elastic wave or some nonlinear wave etc.). We shall take elastic wave equation as an example to describe the MCF method.

The problem to be solved is stated in ref.[1]:
To find $\lambda(x)$, $\mu(x)$ and $\rho(x)$ for $x \in \Omega$ such that the solution of

---

* Institute of Mechanics, Academia Sinica, 100080 Beijing, CHINA.
** Dalian University of Technology, 116024 Dalian, CHINA.

$$(1) \begin{cases} -\sigma_{ij,j}(u) + \rho\dfrac{\partial^2 u_i}{\partial t^2} = 0 \; ; \qquad (x,t)\in\Omega\times(0,+\infty) \\[2mm] \sigma_{ij}(u) = 2\mu\varepsilon_{ij}(u) + \lambda\delta_{ij}\varepsilon_{kk}(u) \\[2mm] \varepsilon_{ij}(u) = \dfrac{1}{2}(u_{i,j} + u_{j,i}) \\[2mm] u = g \; ; \quad x \in \Gamma_D \\[2mm] \sigma_{ij}(u)\cdot n_j = h_i \; ; \quad x \in \Gamma_N \\[2mm] u = 0 \; ; \quad t = 0 \\[2mm] \dfrac{\partial u}{\partial t} = 0 \; ; \quad t = 0 \end{cases}$$

satisfy

$$(2) \qquad Bu = a^* \; ; \qquad x \in \gamma$$

where Ω is a subset of $\mathbb{R}^n$ ($n = 1,2$ or 3) and B is a differential operator with respect to time $t(B_1 u = u,\ B_2 u = \partial u/\partial t,\ B_3 u = \partial^2 u/\partial t^2$, B will be the combination of B_1, B_2 and B_3), and $\Gamma_D \cup \Gamma_N = \partial\Omega$, $\gamma \in \Gamma_N$. And a^* is called the measured data, h and g are known functions.

OBSERVATION

Many inversion methods are more or less presented as optimization problems.(Some examples are given in ref.[2]).They are always associated with one cost-functional which is usually in the form:

$$(*) \qquad J(\lambda,\mu,\rho) = \int_0^T \int_\gamma w\cdot(Bu - a^*)^2 \, d\sigma \, dt \; + \begin{array}{l}\text{regularisation} \\ \text{or penalisation}\end{array} \text{etc}$$

We have done a lot of numerical experiments with functionals as this type.But they do not work well for the points far from source and measurement points.After some observation we find that:
there is a great influence on J for small time data. the large time data have almost no effect on J. as a consequence J is not sensible for λ,μ,ρ far from the sources and receivers.

IDEA OF THE MCF METHOD

The idea of the MCF method is:
using the informations coming from the neighborhood of $x \in \Omega$ to determine the properties λ,μ,ρ at x.

We use the traveltime informations to construct the functionals in order to have as much as possible the information about λ,μ,ρ at a given point.

remark 0. *The advantage of the traveltime inversion method is that it describes clearly the interface of two different materials,while the general optimization method often fails for this.*

In the following we will describe the MCF method for solving the problem (1) and (2) in the time-domain.

FORMULATION OF THE MCF METHOD

Let us, at first, introduce some definitions and notations. For simplicity we suppose that we have a point excitation at the boundary and we set the origin of the coordinate system at this point. Then the boundary conditions may be written as

$$
(3) \quad
\begin{cases}
u = g ; & x \in \Gamma_D \\
\sigma_{ij}(u) \cdot n_j = h_i(t) \cdot \delta(x); & x \in \Gamma_N
\end{cases}
$$

definition 1. *Suppose that* $x \in \Omega$, $x_m \in \gamma$ *and that* $t_x^{x_m}$ *is the travelling time of waves from source to x and from x to x_m (subscript m means measured). We define that a centered influence interval associated to x and x_m is a neighborhood of $t_x^{x_m}$. It will generally be denoted by $I_x^{x_m}$.*

remark 1. *The travelling time of P-waves and S-waves are different. So the centered influence interval for x and x_m is decomposed by several subintervals according to the dimension of the problem. Each subinterval will be used to determine one parameter (e.g. P-wave for Young's moduli, S-wave for shear moduli, etc.). For simplicity and without losing generality we will not distinguish them in the following.*

remark 2. *The choice of $I_x^{x_m}$ for x and x_m is relatively arbitrary. One can choose it according to the excitation function $h(t)$ and the numerical discretization. We will denote $I_x^{x_m}$ by*

$$
(4) \quad I_x^{x_m} = (t_{x,x_m}^1 , t_{x,x_m}^2)
$$

definition 2. *For $x \in \Omega$ we define*

$$
(5) \quad J_x(\lambda,\mu,\rho) = \int_{\gamma_x} \int_{I_x^{x_m}} w \cdot (Bu - a^*)^2 \, dt \, dx_m
$$

where w is a weighting function, $\gamma_x \subseteq \gamma$, and u is the solution of (1).

remark 3. *It should be noted that for a uniform region we need only construct one cost-functional for this region.*

definition 3. *for $x \in \Omega$ we define*

$$
(6) \quad U_x = \left\{ (\delta\lambda, \delta\mu, \delta\rho) \;\middle|\; \begin{array}{l} \delta\lambda(z) = \delta\mu(z) = \delta\rho(z) = 0 \text{ for } z \\ \text{outside a certain neighborhood of } x \end{array} \right\}
$$

definition 4. *For $x \in \Omega$ and (λ,μ,ρ) fixed, we put*

$$
(7) \quad j_x(\lambda,\mu,\rho) = \min_{(\delta\lambda, \delta\mu, \delta\rho) \,\in\, U_x} J_x(\lambda+\delta\lambda, \mu+\delta\mu, \rho+\delta\rho)
$$

Then the MCF formulation is the following

$$(8) \quad \left\{ \begin{array}{l} \textit{To find } \lambda, \ \mu, \textit{ and } \rho \textit{ such that} \\[6pt] j_x(\lambda,\mu,\rho) = J_x(\lambda,\mu,\rho) \\[6pt] \textit{for all } x \in \Omega \end{array} \right.$$

STRATEGY FOR NUMERICAL REALIZATION

An iterative procedure is proposed for the numerical realization of the MCF method:

1> At first we choose a guessed value for (λ,μ,ρ).

2> Start for an x in the neighborhood of the excitation point:

2-1. choose $x_m \in \gamma$ in order to get the major information of $(\lambda(x),\mu(x),\rho(x))$ from the measured data.

2-2. determine $I_x^{x_m}$ with (λ,μ,ρ) by traveltime method.

2-3. modify (λ,μ,ρ) by (5),(7) and (8) using the gradient method (or other optimization methods). then go step by step for x's away from the excitation point.

3> Take the modified (λ,μ,ρ) as guessed value, repeat until (8) is satisfied with a given accuracy.

THE GRADIENT OF $J_x(\lambda,\mu,\rho)$

definition 5. *For $x \in \Omega$ we define*

$$(9) \quad J'_x = \frac{d}{d\alpha}J_x(\lambda+\alpha\delta\lambda, \mu+\alpha\delta\mu, \rho+\alpha\delta\rho)\Big|_{\alpha=0}$$

where $(\delta\lambda,\delta\mu,\delta\rho) \in U_x$.

definition 6. *For $x \in \Omega$, $x_m \in \gamma$ and u being solution of (1) we define the dual state of u associated with the centered influence interval $I_x^{x_m}$ as the solution of*

$$(10) \quad \left\{ \begin{array}{l} -\sigma_{ij,j}(p) + \rho\dfrac{\partial^2 p_i}{\partial t^2} = 0 \ ; \ (x,t) \ \in\Omega\times(0, t^2_{x,x_m}) \\[10pt] \sigma_{ij}(p) = 2\mu\varepsilon_{ij}(p) + \lambda\delta_{ij}\varepsilon_{kk}(p) \\[10pt] \varepsilon_{ij}(p) = \dfrac{1}{2}(p_{i,j} + p_{j,i}) \\[10pt] p = 0 \ ; \quad x \in \Gamma_D \\[10pt] \sigma_{ij}(p)\cdot n_j = B^*(w(Bu-a^*)) \quad \text{for} \quad t>t^1_{x,x_m} \ \text{and} \ x \in \gamma_x \\[6pt] \sigma_{ij}(p)\cdot n_j = 0 \ \text{for other t or other } x \in \Gamma_N \\[10pt] p = 0 \ ; \quad t = t^2_{x,x_m} \\[10pt] \dfrac{\partial p}{\partial t} = 0 \ ; \quad t = t^2_{x,x_m} \end{array} \right.$$

where B^* is the adjoint operator of B.

remark 4. *The only difference between the dual state which is introduced here and the traditional dual state is that here we need $\sigma_{ij}(p)\cdot n_j = 0$ for $t < t^1_{x,x_m}$ or $x \in \Gamma_N\backslash\gamma_x$.*

proposition 1. *For $x \in \Omega$ and fixed $I_x^{x_m}$, we have*

(11)
$$J'_x = \int_\Omega G_{x,\lambda} \cdot \delta\lambda + G_{x,\mu} \cdot \delta\mu + G_{x,\rho} \cdot \delta\rho$$

where

(12)
$$\begin{cases} G_{x,\lambda} =- \int_0^{t^2_{x,Xm}} \varepsilon_{kk}(u) \cdot \varepsilon_{kk}(p) \cdot dt \\[2em] G_{x,\mu} =- \int_0^{t^2_{x,Xm}} \varepsilon_{ij}(u) \cdot \varepsilon_{ij}(p) \cdot dt \\[2em] G_{x,\rho} =- \int_0^{t^2_{x,Xm}} \dfrac{\partial^2 u}{\partial t^2} \cdot p \cdot dt \end{cases}$$

proof. *By (9) we have*

$$J'_x = \int_{\gamma_x} \int_{I^{Xm}_x} 2 \cdot w \cdot (Bu-a^*) \cdot Bu'$$

multiply (10) by u' and integrate over $\Omega x(0, t^2_{x,Xm})$ *we have*

(a)
$$\int_\Omega \int_0^{t^2_{x,Xm}} \sigma_{ij}(p) \cdot \varepsilon_{ij}(u') + \rho\frac{\partial^2 p}{\partial t^2} \cdot u' = \int_{\gamma_x} \int_{I^{Xm}_x} 2 \cdot w \cdot (Bu-a^*)Bu' = J'_x$$

Now, take the variation of (1) with respect to (λ, μ, ρ) *then multiply by p and integrate over* $\Omega x(0, t^2_{x,Xm})$ *we have*

(b)
$$\int_\Omega \int_0^{t^2_{x,Xm}} \sigma_{ij}(p)\varepsilon_{ij}(u') + \rho\frac{\partial^2 u'}{\partial t^2}p + \delta\lambda\varepsilon_{kk}(u)\varepsilon_{kk}(p) + \delta\mu\varepsilon_{ij}(u)\varepsilon_{ij}(p) + \delta\rho\frac{\partial^2 u}{\partial t^2}p = 0$$

and finally with the conditions on u in (1) and on p in (10) we arrive at (11).

remark 5. *The gradient we give here is a global gradient. When we use it in the MCF method we should project it to* U_x.

NUMERICAL TESTS

We have programmed the MCF method in the time-domain for a one dimensional problem. The example is a bar with several different propagation speed of waves at different parts. We suppose that one end of the bar is fixed and the other end is subjected to an impulsive excitation (i.e. a Dirac's measure). We take the velocity (the results are almost the same when we take acceleration) at the excitation end as measured data. The results is given in the following figures. In all figures the vertical axis describes   Young's moduli and the horizontal axis describes the position in the bar referred to the excitation end. We take the length of the bar as 20 with five different layers. The calculation was performed by using the finite element method. In the example 1 (Fig.1) each layer is divided to five elements and in the example 2 (Fig.2)each layer is regarded as one element. In the example 2 we gave several numerical tests when there exist noises (random noise with Gauss distribution) (Fig.3 - Fig.5).

Many other numerical experements show that the converged Young's moduli distribution is not depend on the choice of guessed value. But the speed of convergence will be lower when the guessed value is far from the exact profile.

From the figures we see that for the second iteration the calculated value is almost the same to the exact value when there is no noise. The relative error is not more than 1.0×10^{-3} for second iteration and 1.0×10^{-5} for 5th iteration.

For 1 percent noise the relative error is about 2.0×10^{-2}, for 5 percent noise the relative error is about 4.0×10^{-2} and for 10 percent noise the relative error is about 3.0×10^{-1}.

DISCUSSION

From the numerical experiments it appears that the formulation of the MCF method given by (8) is well posed. The example we have taken has a big jump on Young's moduli (this is very frequency for geophysical problems),for this conventional methods do not give good result.The comparison with other methods is that the present method has a simple physical interpretation, a better convergence and a better stability. And also it can be easily adapted to 2 or 3 dimensional problems.There are two orientation for multi-dimensional problems: (1.take a approximate evaluation of the distribution of parameters with other method,like traveltime method,migration etc.,before performing the MCF method. 2.adapt directly the MCF method).

It should also be pointed out that this (the MCF) method can be used to analyze local structures.

REFERENCES

[1]. DING H., ZHENG Z.M., XU S.Z., *A new approach to inverse problems of wave equations (I)*. Appl. Math. and Mech. vol.11, n°12(1990), pp.1043-1047.
[2]. TARANTOLA,A., *Inverse problem theory*. Elsevier, 1987.

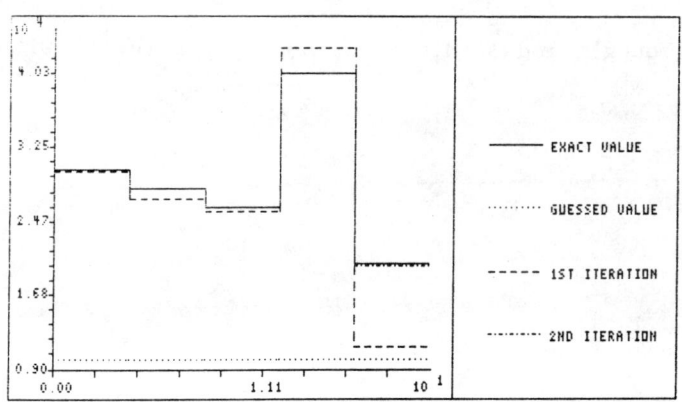

Fig.1 Young's moduli distribution / Depth (e.g.1 without noise)

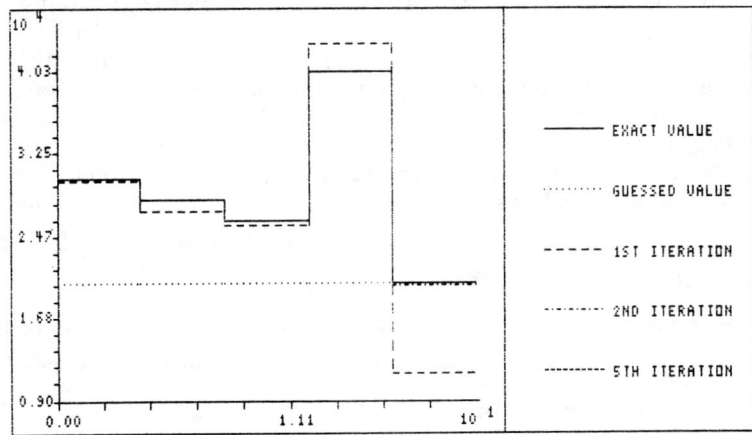

Fig.2 Young's moduli distribution / Depth (e.g.2 without noise)

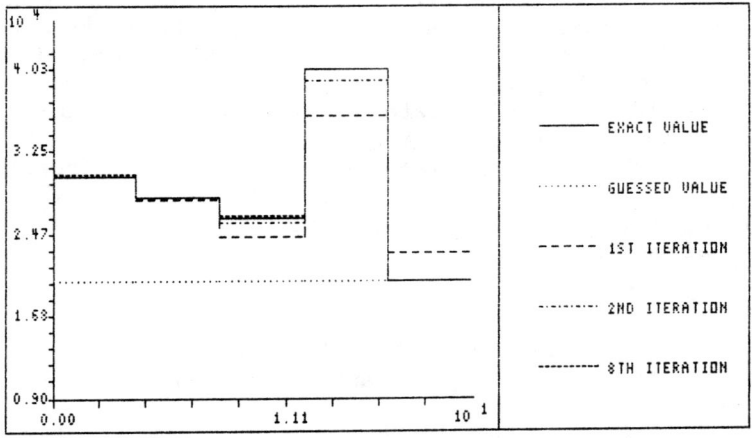

Fig.3 Young's moduli distribution / Depth (e.g.2 with 1% noise)

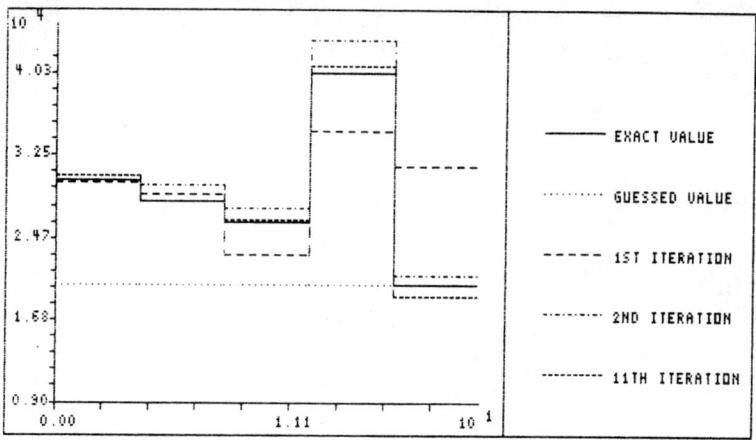

Fig. 4 Young's moduli distribution / Depth (e.g. 2 with 5% noise)

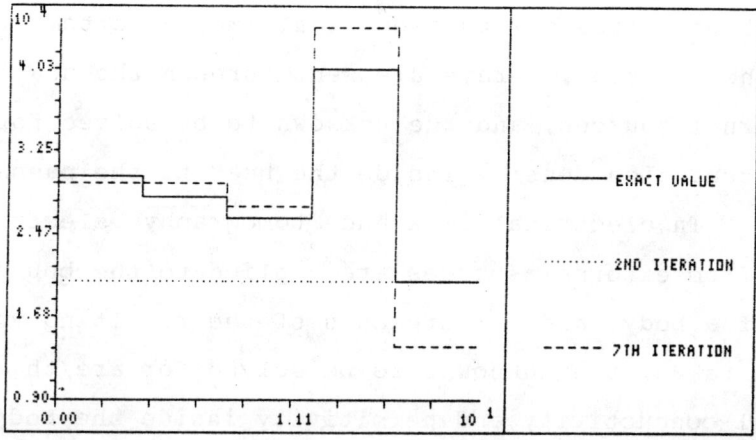

Fig. 5 Young's moduli distribution / Depth (e.g. 2 with 10% noise)

CHAPTER 3

The CIT SCAN

C. Denson Hill*
Robert D. Sidman†

1. <u>Introduction</u>. The CIT scan for EEG is short for the
<u>Cortical</u> <u>Imaging</u> <u>Technique</u> in electroencephalography. As
a diagnostic tool for medical applications it is intended
to be complimentary to the more familiar CAT scan. It
should not be confused with electrical impedance tomography. In the CAT scan, x-rays are sent through the head
from external sources, and the unknown to be solved for
is the attenuation density inside the head to the passage
of x-rays. In electrical impedance tomography, electric
currents from external sources are applied to the boundary
surface of a body, and measurements of the resulting voltages are taken; the unknowns to be solved for are the
electrical conductivity and permitivity inside the body.
In the CIT scan, no external currents or voltages are ap-
lied; what is being measured is the electrical activity
produced by the brain itself ("brain waves"). The purpose
of this article is to give a clear conceptual presentation
of the Cortical Imaging Technique, and to explain to what
extent it addresses the fundamental inverse problem of
electroencephalography.

(*) Dept. of Mathematics, SUNY at Stony Brook, Stony Brook,
    N.Y. 11794.

(†) Dept. of Mathematics, University of Southwestern Louisiana,
    Lafayette, LA 70504.

The CIT scan has the character of what is usually called an inverse problem in partial differential equations. Its connection with Geophysical Fluid and Solid Mechanics stems from the fact that it was inspired by the work on inverse problems, motivated by geophysical applications, that was done by Jim Douglas Jr. [1],[2] John Cannon [3], Cannon & Douglas [4], and Keith Miller [5]quite some time ago. The CIT method was first reported in the literature by the current authors in Hill et al [6] and Sidman et al [7]. One should also mention Kearfott et al [8], where further references can be found. We would like to acknowledge the collaboration of R. Baker Kearfott, Diana J. Major, Martin R. Ford, Dennis B. Smith, C. Schlichting, Lu Lee and Ronald Kramer.

2. The Basic Problem. In electroencephalography (EEG) sensors are placed on the scalp and the voltages at those known locations are recorded as a function of time. There may be only 16 or even fewer sensors, or up to as many as 28 or 30. To be concrete let us assume for the purposes of this article that there are 28 sensors applied at known locations. Thus the raw scalp data consists of 28 graphs which look something like Figure 1:

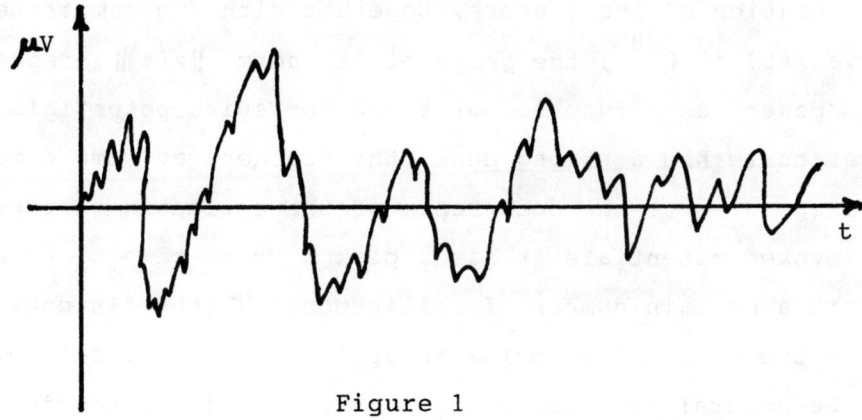

Figure 1

The basic question is: "What do you do with these 28 graphs?"

It should be pointed out that these EEG recordings
are taken in basically two different contexts: I. "Evoked"
potentials & II. "Spike" potentials. Evoked potentials
refers to the attempt to record the electrical activity
in the brain that is evoked in response to certain stan-
dardized external stimuli, such as a mild electric shock
to a thumb, a flashing of light in the eyes, or by putting
various sounds in an ear. Spike potentials refers to the
attempt to record the electrical activity in the brain
during or at the onset of a spontaneously generated epileptic seizure. The evoked potentials are quite weak
($\sim 10^{-6}$ volts) and are almost completely masked by a great
deal of noise of the same order of magnitude as the signal,
due to the constant random electrical activity normally
taking place in the brain. Thus for evoked potentials
periodic stimuli are used, and an average over many trials
is taken so as to average the more or less random noise out
of the signal. What is left is a discernable but still very
noisy response. For spike potentials the signal is dramatic
enough to be discerned in spite of the random noise, without
averaging, but of course much noise is still present in the
raw data.

In any event the raw data corresponds to knowledge of
the location of the sensors, together with a parametrized
curve $x(t)$ in $\mathbb{R}^{28}$; the graph of its norm $\|x(t)\|$ represents
the "power" as a function of time. For spike potentials this
power curve has a strong peak (the "spike") at some time t_0
corresponding to the occurrence of the epileptic seizure.
For evoked potentials it has a peak at some time t_0 which
occurs a certain number of milliseconds τ (the "latency")
after the external stimulus is applied. In practice there
may be several such peaks, with corresponding latencies, but
for our purposes here we will concentrate on just one of them.

3. Possible Approaches. One approach is to throw away 27
of the graphs, keeping only the one that corresponds to
the sensor located nearest the region of expected brain
activity. From this one graph one usually extracts only
two numbers: the latency τ and the amplitude of that
graph at it's power peak. Thus from a parametrized curve
$x(t)$ in $\mathbf{R}^{28}$ one extracts only two numbers. Such a crude
approach is actually effective in studying the demyelina-
tion of nerves, where an increased latency and reduced
amplitude would be observed.

A second method is to fix the time t_o and look at
the 28 numbers $x(t_o)$ as representing a discrete graph of
the voltage profile ploted over the scalp. By piecewise
linear interpolation one then gets a continuous graph of
the voltage profile on the scalp. Typically a contour map
of the various level curves of this interpolated graph is
then drawn, and the various levels are painted in pretty
colors. These are the pictures one sees coming out of a
"brain imaging" machine. Aside from it's inherent crudity
this method has a serious drawback: The high resistivity of
the skull and other material between the brain and the
scalp greatly attenuates the signal; what one sees on the
scalp is a very smoothed and damped version of what the
voltage profile might actually be, for example, on the
cerebral cortex.

A third approach is to try to determine, at the fixed
time t_o, an equivalent single dipole source located at some
unknown position, at some unknown orientation and strength,
somewhere inside the brain. This is the "dipole localization
method" (DLM) due to Sidman et al [9],[10],[11],[12]. It
is determined by finding a dipole whose potential field
gives the best least squares fit to the scalp data at the 28

sensor locations. This is a nonlinear problem involving
6 free parameters which must be adjusted to fit the 28
data points. For a discussion of it's effectiveness and
utility, see the references mentioned above. There are
two main limitations of the DLM approach: For some sets
of data it is impossible to obtain a good fit by a single
equivalent dipole source, and by it's very nature it does
not distinguish between a single source and multiple
sources. If one tries to employ several unknown dipoles,
one runs into problems of nonuniqueness in the solution
of an unwieldy nonlinear problem.

4. The CIT scan. The new approach presented here, the
Cortical Imaging Technique, also uses the 28 scalp data
points at the fixed time t_o, but calculates a simulated
potential field right at or very near the actual cortical
surface. This avoids the difficulties of an attenuated
signal mentioned in the second approach above, and is capable in many cases of separating and detecting multiple
sources, even when they may be deep inside the brain.

In what follows we try to give a conceptual description of the essence of the CIT approach, using a minimum
of mathematical formulas (for a more detailed but less
conceptual discussion see [6],[7],[8]): Suppose we model
the head by a homogeneous conducting ball of unit radius,
and the cerebral cortex by the upper half of a concentric
hemisphere H of radius R < 1.

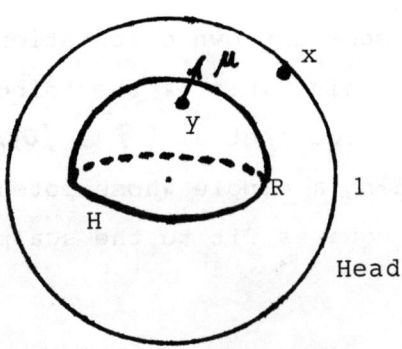

Figure 2

On H we place an unknown dipole moment density μ. The resulting "double layer" potential V is given by

(1)
$$V(x) = \int_H k(x,y)\mu(y)d\sigma(y) ,$$

where

$$k(x,y) = \frac{\partial}{\partial n}(\frac{1}{r}) ,$$

$r = \|x-y\|$, n is the outer unit normal to H, and $d\sigma$ represents the surface measure on the hemisphere. (In actual practice the head is modeled by a more complicated threelayer system of spheres of different conductivities, and a much more complicated formula for V(x) is used, due to Wilson and Bayley [13], which more accurately gives the potential field induced by a dipole inside an equivalent homogeneous conducting ball; see [7],[8].)

Next we replace (1) by an appropriate approximating Riemann sum using N small subdivisions of H, with N >> 28. So now we have a finite number of unknowns which form a vector $\vec{\mu} = (\mu_1, \mu_2, \ldots \mu_N)$ in $\mathbb{R}^N$. We require that the discretized potential V(x) should agree with the observed experimental data at the 28 known sensor locations. Note that now we have, in effect, a system of N small dipoles, each pointing radially outward and distributed over the hemispherical surface H, of some unknown strength $\{\mu_i\}$. Thus we arrive at a <u>linear</u> <u>system</u> of 28 equations for N unknowns $\vec{\mu}$. There are of course many solutions because it is highly underdetermined.

If we look for the <u>least</u> <u>energy</u> <u>configuration</u>, in the discrete approximation, it means that we must find the unique solution which has minimum L^2 norm in $\mathbb{R}^N$:

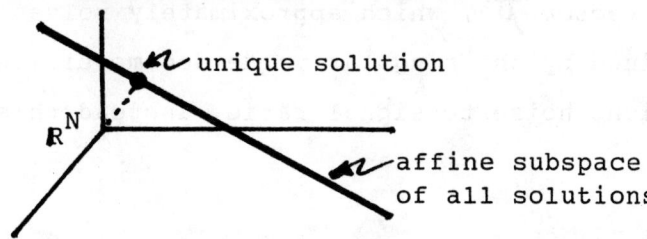

Figure 3

In other words, we are looking at the solution $\vec{\mu}$ to the quadratic programing problem

(2) $\begin{cases} A\vec{\mu} = \vec{b} & \text{(28 by N system)} \\ \|\vec{\mu}\| = \text{minimum} & (L^2 \text{ norm}) \end{cases}$.

However problem (2) as it stands may be quite ill conditioned. In particular the inhomogeneous term $\vec{b}$ comes from the experimental data, which contains a very high percentage of noise. For this reason we use the singular value decomposition [14], which allows us to "tune" the problem by some realistic "noise to signal ratio" (say, .05 for example). Given an estimate ε of the relative errors in the data, we may effectively replace the system (2) by a system having possibly lower rank so that errors of size ε will not unduly influence the solution:

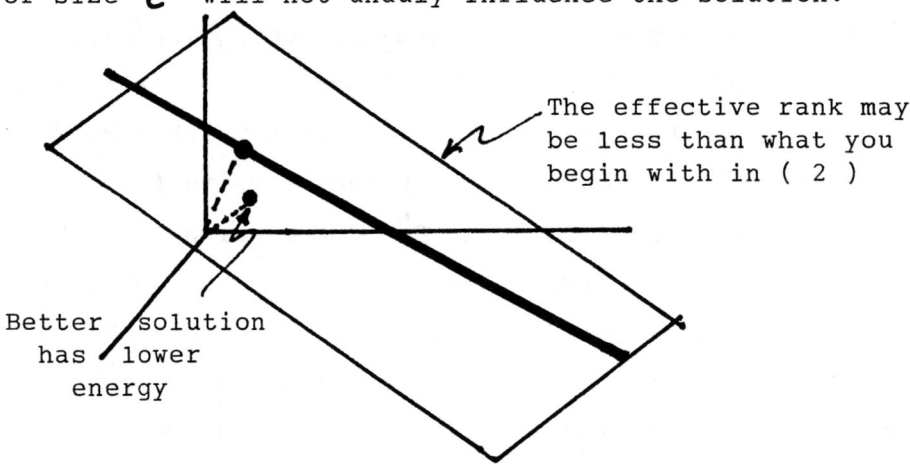

The effective rank may be less than what you begin with in (2)

Better solution has lower energy

Figure 4

Thus the 28 by N linear system in (2) is not to be solved exactly, but rather to within a certain tolerance ε . This will ultimately lead to a final answer which is more regular with fewer artifactual bumps.

Our final answer does not however consist of the solution vector $\vec{\mu}$, which approximately solves (2) and is determined by the singular value decomposition using some convenient noise to signal ratio. Instead this $\vec{\mu}$ is inserted

into the discrete Riemann sum approximation to (1),
and the resulting potential V(x) is ploted, as a graph
over the cerebral cortex, using the values of V(x) for
$\| x \| = R + \delta$, with $\delta > 0$ very small:

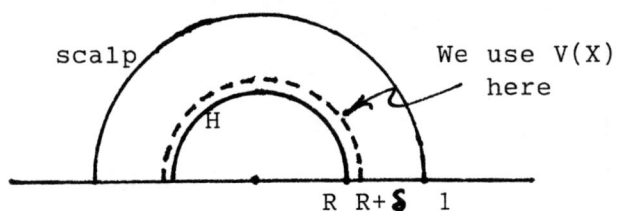

Figure 5

The reason for the δ can best be understood by studying
the works of Cannon, Douglas, and Miller cited in the
introduction. In practice there may be some tuning of
the choice of δ in conjunction with the choice of the
noise to signal ratio ε.

We now turn to some numerical and experimental veri-
fications of the method. We must be brief for lack of space.

5. <u>Separation of sources</u>. First we perform a numerical
experiment to see if we can match results with know arti-
fically concocted data: We put in two rather deep dipole
sources of known strengths as shown in Figure 6. Here we
are using the threelayer model; X denotes the right ear,
Y the nose, and Z the top of the head.

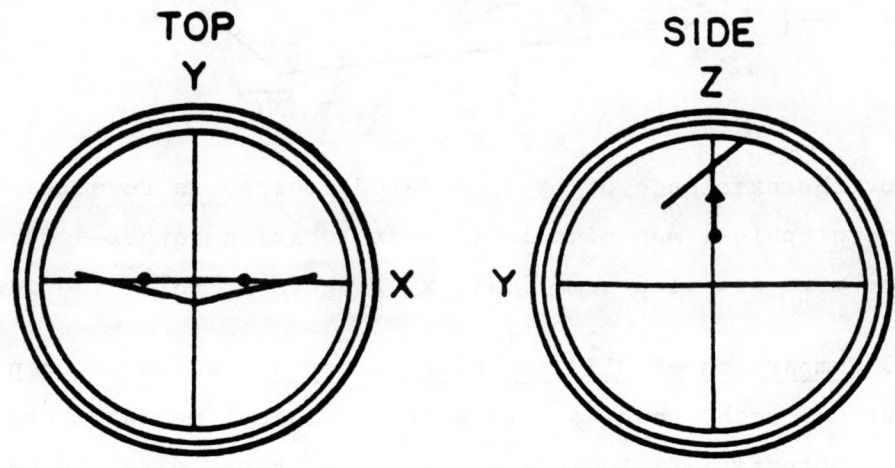

Figure 6

The exact potential field on the scalp can be found using the formulas of Wilson and Bayley, and is shown in Figure 7 in which the scalp has been flattened into a rectangle:

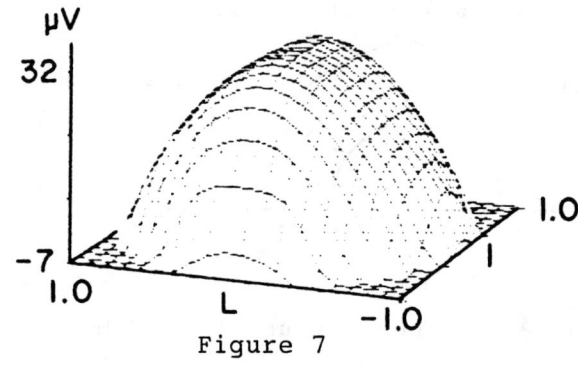

Figure 7

Notice the pair of sources is not discernable even from the exact scalp data. Next we sample the exact scalp data at 28 standard locations, add 10% randomly generated noise, and use the noisy data in the CIT scan to obtain the field on the simulated cerebral cortex:

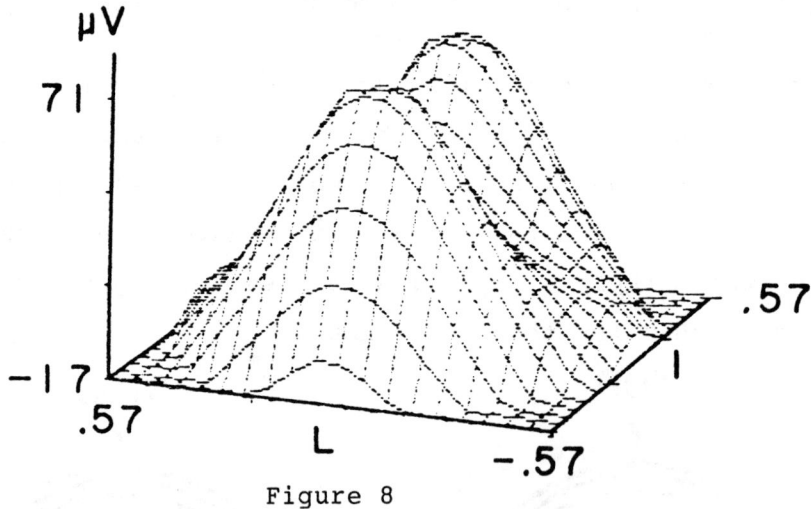

Figure 8

Now the existence of two separated sources is revealed; a topographical map pinpoints their location rather accurately. For more detailed numerical experiments of this kind see [8].

6. Comparison of CIT with clinical data. Second we compare our CIT scan results with actual clinical measurements of the potential field made by placing sensors directly on the

cerebral cortex. The experiment involves the response in the left central-temporal region evoked by right median nerve stimulation (a mild electric shock to the right thumb). This clinical data was provided to us by courtesy of Dennis Smith, M.D. of the Oregon Comprehensive Epilepsy Program at Good Samaritan Hospital, Portland, Oregon.

Before surgery scalp data was taken and our CIT scan method was used to generate a topographical map of the field on the simulated cerebral cortex. The left central-temporal portion of it is shown in Figure 9A, in which the right peak is at level 16, and the left valley at -2.4. Later direct measurements were taken for that region using an 8 x 8 cortical patch, and a contour map of the field was ploted. It is shown in Figure 9B, in which the right peak is at level 17.7 and the left valley at -6.9.

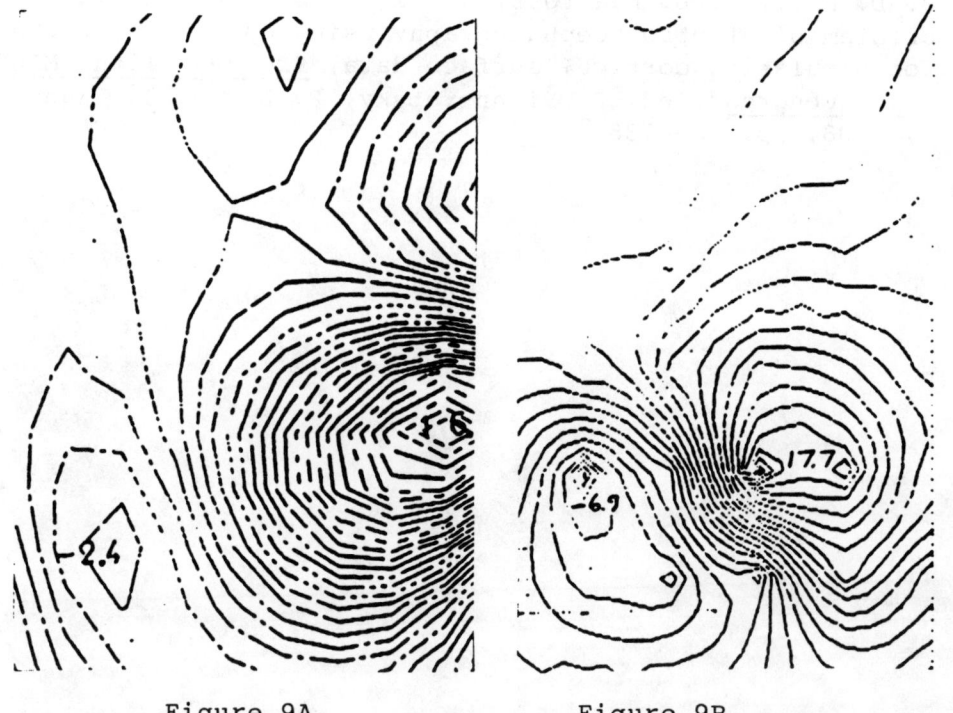

Figure 9A   Figure 9B

All values are in microvolts. The results appear to be in remarkable agreement, considering that it was impossible to perform the two experiments simultaneously.

REFERENCES

1. J. Douglas, Mathematical programing and integral equations, Symposium Provisional International Computation Center, Birkhäuser Verlag, Basel, 1960.

2. J. Douglas, A numerical method for analytic continuation, in Boundary Value Problems in Differential Equations, University of Wisconsin Press, pp. 179-189, 1960.

3. J. R. Cannon, Some numerical results for the solution of the heat equation backward in time, in Numerical Solutions of Nonlinear Differential Equations, ed. D. Greenspan, John Wiley & Sons, 1967, pp.21-54.

4. J. R. Cannon and J. Douglas, The approximation of harmonic and parabolic functions on half-spaces from interior data, CIME 2º Ciclo, Edizioni Cremonese, 1967, pp. 193-230.

5. K. Miller, Least squares methods for ill-posed problems with a prescribed bound, SIAM J. Math. Anal. 1, 1970, 52-74.

6. C. D. Hill, R. B. Kearfott, and R. D. Sidman, The inverse problem of electroencephalography using an imaging technique for simulating cortical surface data, Proceed. 12th IMACS World Congress, ed. R. Vichnevetsky, P. Borne, J. Vignes, 3, 1988, pp. 735-738.

7. R. D. Sidman, R. B. Kearfott, D. J. Major, C. D. Hill, M. R. Ford, D. B. Smith, L. Lee, and R. Kramer, Development and application of mathematical techniques for the non-invasive localization of the sources of scalp-recorded electric potentials, in Biomedical Systems Modelling and Simulation, Vol 5 IMACS Trans. Scientific Computing, ed. J. Eisenfeld and D. S. Levine, J. C. Baltzer AG, Basel, 1989, pp. 133-157.

8. R. B. Kearfott, R. D. Sidman, D. J. Major, and C. D. Hill, Numerical tests of a method for simulating electrical potentials on the cortical surface, to appear in IEEE Trans. Biomedical Eng.

9. R. D. Sidman, V. Giambalvo, T. Allison, and P. Berry, A method for localization of sources of human cerebral potentials evoked by sensory stimuli, Sensory Processes 2, 1978, pp. 116-129.

10. R. D. Sidman, The time-dependent equivalent dipole source for the response to median nerve stimulation, IEEE Trans. Biomedical Eng., 31, 6, 1984, pp. 481-483.

11. D. B. Smith, R. D. Sidman, J. S. Henke, D. Labiner, and H. Flanigin, A reliable method for localizing deep intra-cranial sources of the EEG, Neurology, 35,12, 1985,1702-1707.

12. L. Lee, D. B. Smith, R. D. Sidman, and R. Kramer, Intra-cranial localization of epileptic spikes using DLM, J. Clin. Neurophysiol., 5, 4, 1988, 336 (abstract).

13. F. N. Wilson and R. H. Bayley, The electric field of an eccentric dipole in a homogeneous spherical conducting medium, Circulation, 1, 1950, pp. 84-92.

14. G. W. Stewart, Introduction to Matrix Computations, Academic Press, New York, 1973, pp. 317-326.

CHAPTER 4

On Uniqueness of the Discontinuous Conductivity Coefficient

Victor Isakov*

Abstract. We give recent results about uniqueness of the coefficient a of the equation div ($a \nabla u$) = 0 with given boundary Cauchy data. In case of one set of boundary data we assume a is the sum of unity and of the indicator function of an unknown domain.

There are many important inverse problems concerning shape determination. We mention the inverse problem of potential theory, the inverse scattering problem and the inverse seismic problem in the case when the density coefficient is piecewice smooth. These problems simulate search of unknown objects. They are non-linear and, as a rule, ill-posed. So theoretically, uniqueness results are of most importance here.

Recently, a significant amount of attention has been attracted to the so-called inverse conductivity problem where one looks for the coefficient a of the elliptic equation

$$div\,(a \nabla u\,) = 0$$

* Department of Mathematics and Statistics, Wichita State University, Wichita, KS 67208

given the results of many (the Neumann-to-Dirichlet map) or single boundary measurements. This problem has applications in any case in medicine (electrical tomography) and geophysics (magnetic prospecting). Starting with ideas of A. Calderon, John Sylvester and Gunther Uhlmann in the remarkable paper [21] completely resolved the uniqueness question in the three-dimensional case when a is smooth enough. There are certain generalizations and many applications of their results; for details we refer to the reviews of the author [12] and of Sylvester and Uhlmann [22]. At present, there are no uniqueness results for, say, Lipschitz a, so discontinuous coefficients create certain difficulties. Observe that in applications such coefficients are quite common.

In this lecture we describe some uniqueness reults for discontinuous a in case of many and single boundary measurements and we give some applications to the inverse scattering problem (inverse transmission problem). We consider a of the form $a_0 + b \, \chi(D)$ where a_0 is a given fuction, b is an unknown function and $\chi(D)$ is the characteristic function of an unknown open set D.

Many boundary measurements.

Let Ω be a bounded domain in $\mathbb{R}^n$, $n = 2, 3$, with the C^2 – boundary. Consider the   Dirichlet problem

$$div \, ((a_0 + b_j \, \chi(D_j)) \, \nabla \, u_j \,) = 0 \qquad \text{on } \Omega$$

$$u_j = g_0 \qquad \text{on} \qquad \partial\Omega$$

(1)

where a_0, b belong to C^2 ($cl \, \Omega$), $0 < \epsilon < a_0 + b_j$ for certain number ϵ. We assume that D_j is an open subset of Ω with $cl \, D_j \subset \Omega$. It is well-known that in this case there is a unique solution u_j ($; g_0$) to the problem (1) in the Sobolev space $H_{(1)}(\Omega)$ which is a C^1 – function on the closure of some neighborhood of $\partial\Omega$ in Ω provided g_0 belongs to $C^2(\partial\Omega)$.

Let Γ be a non-void open subset of $\partial\Omega$.

<u>Theorem 1.</u>   Suppose D_j have the Lipschitz boundaries and the complements $\Omega \setminus cl \, D_j$ are connected. Suppose $b_j \neq 0$ on $\partial \, D_j$. If

$$\partial \, u_1 \, (\; ; g_0 \,) \, /\partial N \; = \; \partial \, u_2 \, (\; ; g_0 \,) \, /\partial \, N \qquad \text{on } \Gamma$$

for all $g_0 \in C^2 (\partial \Omega)$ which are zero outside of Γ then $D_1 = D_2$. If $n = 3$ then in addition $b_1 = b_2$ on D_1.

We denote by N the exterior unit normal to $\partial \Omega$.

Observe that since we are given the normal derivative for all Dirichlet data we have physically many boundary measurements.

A proof of this result is given in the author's paper [8]. We explain its basic idea. Subtracting two equations (1) and letting $u = u_2 - u_1$ and transferring the term containing the difference of the coefficients into the right side we get

$$div \left((a_0 + b_2 \, \chi(D_2)) \nabla u \right) = div \left((b_1 \, \chi(D_1) - b_2 \, \chi(D_2)) \nabla u_1 \right) \quad \text{in } \Omega$$

Since the Cauchy data for u are zero on Γ we conclude that $u = 0$ outside the simply connected hull of $D_1 \cup D_2$. So using the definition of a weak solution we have that the right side is orthogonal to all solutions to the adjoint equation, or

$$\int_{D_1} b_1 \, \nabla \, u_1 \cdot \nabla v_2 = \int_{D_2} b_2 \, \nabla \, u_1 \cdot \nabla v_2 \tag{2}$$

for all solutions v_2 to the equation (1) with $j = 2$ near the closure of the mentioned hull. By using the Runge property we transfer these relations onto all solutions v_1 near the same hull. If D_1 does not belong to D_2 then there is a point z in $\partial D_1 \setminus cl \, D_2$. Let $v_j(x) = K_j(x, y)$ where K_j is a fundamental solution to the equation (1) and y is a point outside $cl \, (D_1 \cup D_2)$. If y tends to z the integral in the left side of the equality (2) is unbounded, for the integrand behaves as $|x - y|^{2-2n}$ while the integral in the right side is bounded. This leads to a contradiction and shows that $D_1 \subset D_2$. Due to the symmetry we get the inverse inclusion and therefore the equality of D_1 and D_2.

We observe that Kohn and Vogelius [14] proved that when a_0, b_j and ∂D_j are analytic it is possible to identify also a_0, in fact they proved uniqueness of any piecewise analytic conductivity coefficient a with a given Dirichlet-to-Neumann map. As for the smooth case, recently Jeff Powell showed that the condition $b_j \neq 0$ on ∂D_j and the condition $cl \, D_j \subset \Omega$ can be relaxed. It looks however that there are no uniqueness results for general piecewise smooth a. We think that in this case still there is uniqueness.

Now we give an application of the method described to the inverse transmission scattering problem. We observe that there is an intimate connection between the Dirichlet-to-Neumann map and the scattering amplitude, and we refer about details to the paper of Nachman [20] and already mentioned review paper of Sylvester and Uhlmann [22].

Let D be a bounded open set in $\mathbb{R}^n$ with the Lipschitz boundary. Let D^e be $\mathbb{R}^n \setminus cl\, D$ and let D^i be D. Let a function μ belong to $C^2 (\mathbb{R}^n)$, $\mu > 0$ on $\mathbb{R}^n$, and let a function ρ belong to $L_\infty (\mathbb{R}^n)$. A transmission scattering problem is to find functions u^e, u^i on D^e, D^i satifying the following differential equations

$$\Delta u^e + k^2 u^e = 0 \quad \text{on} \quad D^e$$

$$\text{(3)}$$

$$div\,(\mu \nabla u^i) + k^2 \rho\, u^i = 0 \quad \text{on} \quad D^i$$

the transmission boundary conditions

$$u^e = u^i, \qquad \partial u^e / \partial N = \mu\, \partial u^i / \partial N \qquad \text{on } \partial D \qquad \text{(4)}$$

and the condition at infinity

$$u^e (x) = exp (i\, x \cdot \xi) + u^e_0 (x), \qquad |\xi| = k \qquad \text{(5)}$$

where u^e_0 satisfies the radiation condition

$$lim\; r^{(n-1)/2} (\partial u^e_0 / \partial r - i\, k\, u^e_0) = 0 \quad \text{as } r = |x| \text{ tends to } +\infty \qquad \text{(6)}$$

Observe that if $\mu = \rho$ is constant we obtain the transmission scattering problem considered by Colton and Cress [4] and by Angell, Kleinman and Roach [2] where there are existence and uniqueness theorems for this direct scattering problem in the case $\partial D \in C^{1+\lambda}$, $0 < \lambda < 1$.

For non-smooth solutions the conditions (4) should be understood in a weak sense as follows. Letting

$$u = u^3 \text{ on } D^3, \; u = u^i \text{ on } D^i, \; a = 1 + (\mu - 1)\, \chi(D), \; c = 1 + (p - 1)\, \chi(D)$$

we rewrite the equations (3) and the transmission condition (4) in the following form

$$div(a \nabla u) + k^2 c\, u = 0 \quad \text{in } \mathbb{R}^n \qquad \text{(7)}$$

In the paper [13] we prove that under the conditions imposed on D, μ, ρ, for any $\xi \in \mathbb{R}^n$, $|\xi| = k$, there is a unique solution u to the equation (7) satisfying the conditions (5), (6). This result seems to be new in such generality, however we observe that it is related to Theorem 2.3 of the book of Lax and Phillips [16]. From the condition (6) and from the well known results it follows that

$$u_0(x) = r^{(1-n)/2} \, exp\,(ikr)\,(A(\sigma, \xi) + 0\,(r^{-1}))$$

as r tends to $+\infty$. Here $\sigma = r^{-1} x$.

The inverse transmission problem is to find D, μ, ρ (or, equivalently, a, c) given the scattering amplitude $A(\sigma, \xi; k)$. We consider the case when A is known for all $\sigma \in \Gamma(1)$, $\xi \in \Gamma(k)$ for one or two values of the frequency k. Here $\Gamma(k) = \{y \in \mathbb{R}^n : |y| = k\}$. This problem is of significance for acoustic and electromagnetic prospecting. It was solved numerically in several papers (e.g. the paper of Angell, Kleinman and Roach [2]), but as noted by Rainer Kress in his lecture in Arcata, California, in August 1989, there are no uniqueness results even for constant μ, ρ.

To formulate our uniqueness theorems we assume that there are two open sets D_1 and D_2 and coefficients μ_1, ρ_1 and μ_2, ρ_2 and we denote the related solutions to the transmission scattering problem by u_1 and u_2 and their scattering amplitudes by A_1 and A_2.

In Theorems 2, 3, 4 we assume that $\mu_j \neq 0$ on ∂D_j and that D_j^e are connected.

Theorem 2. If

$$A_1(\sigma, \xi) = A_2(\sigma, \xi) \tag{8}$$

for all $\sigma \in \Gamma(1)$ and all $\xi \in \Gamma(k)$ for one frequency $k > 0$ then

$$D_1 = D_2 \text{ and } \mu_1 = \mu_2 \text{ on } \partial D_1.$$

In Theorems 3 and 4 we in addition assume that $n = 3$.

Theorem 3. Suppose that the relation (8) is valid for one frequency $k > 0$.
Then i) if $\rho_1 = \rho_2$ then $D_1 = D_2$, $\mu_1 = \mu_2$ on D_1 , ii) if $\mu_1 = \mu_2$ then $D_1 = D_2$, ρ_1
$= \rho_2$ on D_1.

In Theorem 4 we consider the more general equation

$$div(a\nabla u) + (k^2 c + b)\, u = 0 \qquad\qquad \text{in } \mathbb{R}^n$$

instead of the equation (7). We assume that $b = \nu\, \chi\,(D)$ with $\nu \in L_\infty\,(\mathbb{R}^3)$.

<u>Theorem 4.</u> Suppose that the relation (8) is valid for all $\sigma \in \Gamma(1)$ and for all $\xi \in \Gamma(k)$, $k = k_1$ or k_2, $0 < k_1 < k_2$.

Then i) if $\rho_1 = \rho_2$ then $a_1 = a_2$, $b_1 = b_2$, ii) if $\nu_1 = \nu_2$ then $a_1 = a_2$, $b_1 = b_2$ and iii) if $\mu_1 = \mu_2$ then $D_1 = D_2$, $b_1 = b_2$, $c_1 = c_2$.

We observe that in the case of Dirichlet or Neumann data on ∂D instead of the transmission data, uniqueness theorems have been obtained by Lax and Phillips and Schiffer (see the book [16], p. 173) and by Colton and Kress (the book [4], p. 195) if A is given for all frequences and all directions of receivers σ, or for one frequency and for all directions of receivers and all directions of incident waves ξ, while for backward scattering Lax and Phillips [17] and Majda [19] proved uniqueness of the convex hull of the scattered D. A particular case of transmission conditions has been considered by Majda and Taylor in the paper [18] where a convex scatterer was restored from the high-frequency behavior of A.

We give an outline of the proof, say, for Theorem 2.

First we prove that for any open set Ω with connected complement $\mathbb{R}^n \backslash\ cl\ \Omega$ which contains $cl\ D$ any solution to the equation (7) in Ω is the L_2-limit of linear combinations of solutions to the scattering problem (5), (6), (7) on any compact subset of Ω. From the condition (8) it follows that the difference u of two possible solutions to the transmission scattering problems with the coefficients a_2, c_2 and a_1, c_1 is zero outside a large ball. So subtracting the related equations (5) and acting as above in Theorem 1 we conclude that

$$\int_{D_1} (\mu_1 - 1)\, \nabla u_1 \cdot \nabla v_2 + k^2 c_1 u_1 v_2 =$$
$$\int_{D_2} (\mu_2 - 1)\, \nabla u_1 \cdot \nabla v_2 + k^2 c_2 u_1 v_2$$

for any solution v_2 to the equation (5) with $a = a_2$, $c = c_2$. Then using the above remark concerning approximation we transfer this relation onto all solutions v_1 to the first equation. The proof can be completed now as in Theorem 1. Of course, there are many important technical details of this proof which appeared in the paper [13].

Single boundary measurements.

Now we assume that for solutions to the problem (1) the normal derivaties are equal only for one choice of the Dirichlet data g_0. Then we impose additional requirements that $a_0 = b_j = 1$, wo we are looking only for the shape of unknown set D_j. We obtain the ill-posed free boundary problem which is somehow similar to the well-known inverse problem of potential theory. About the latter problem we refer to the author's book [10]. The problem under consideration however seems to be more difficult; at present no global uniqueness results are known except of the simple case when $D_1 \subset D_2$ or when D_j are convex polyhedrons or cylinders. In the first case a proof of uniqueness and a stability estimate can be found in the paper of Alessandrini [1]. Initially this problem has been considered locally by Friedman and Gustafsson [5].

We give precise results for the second case assuming that Ω is either a bounded domain with the C^2 – boundary or a halfspace when in addition we prescribe the natural behavior at infinity as for potentials.

Theorem 5. Consider two cases

a) D_j are convex polyhedrons
b) D_j are bounded cylinders with strictly convex bases.

Suppose that

$$dist\,(D_j\,,\,\partial\Omega) > diam\,D_j \qquad\qquad (9)$$

If for solutions u_j to the Dirichlet problem (1) we have

$$\partial u_1(\quad;g_0)\,/\,\partial N = \partial u_2(\quad;g_0)/\partial N \quad \text{on} \quad \Gamma$$

for some $g_0 \in C^2(\partial\Omega)$ that is not identically zero then in the case a) we have D_1 $= D_2$ and in the case b) we have $D_1 = D_2$ or $D_1 = D_{2\#}$ where $D_{2\#}$ is the cylinder with the same base as D_2 and there is no more than one such $D_{2\#}$.

The proof is given in the paper of Friedman and Isakov [6] (case a) and in the paper of Isakov and Powell [11] (case b). This proof is based on an analysis of singularities of u_j near edges of D_j. It is not clear at the time whether the condition (9) is essential, also it is not clear that actually there is a second solution $D_{2\#}$ in the case b).

A local analysis of this problem has been started in the paper of Bellout and Friedman [3] where it has been observed that a linearization of this problem is related to the oblique derivative problem for the Laplace equation when the direction of the oblique differentiation is tangent to the boundary of Ω. This problem is not elliptic possibly except for $n=2$. In the latter case Bellout, Friedman and Isakov recently obtained some results concerning local uniqueness and stability. They assumed that ∂D_j is analytic while D_j itself is simply connected, then if, say, the function g_0 has only one maximum and one minimum on $\partial\Omega$ then the index of the gradient of u_j is zero, the related oblique derivative problem has always a solution and a solution of the inverse problem is locally unique. The paper is in preparation. Practically, this result suggest that for best computational result one should use functions g_0 with the described properties.

The picture is somewhat obscure even if D_j is the union of disks. We give the following result.

Theorem 6. Let $n = 2$ and let Ω be a half-plane. Suppose that D_j is the union of disks $B_{j1}, \ldots, B_{jk(j)}$ whose closures are disjoint such that the centers of B_{jk}, $k = 1, \ldots, k(j)$, are extreme points of their convex hull. If g_0 is not identically zero and if the normal derivatives of $u_j(\ ; g_0)$ coincide on Γ then D_1 $= D_2$.

For $k(j) = 1$ (i.e. when D_j are disks) this result has been proven by Friedman and Isakov [6] and in the general case a proof has been give by Powell [11]. These proofs discover a quite complicated picture of singularities of the harmonic contunuation of the function u_j from $\Omega \backslash cl\ D_j$ across ∂D_j, they result from reflections of the centers of the disks with respect to $\partial\Omega$ and from subsequent inversions with respect to the boundaries of the disks forming D_j. It means probably that this inverse problem is not simple at all.

For computational and practical aspects of the inverse conductivity problem we refer to the papers of Isaacson [7] and of Kohn and Vogelius [15].

References

1. G. ALESSANDRINI, *Remark on a paper of Bellout and Friedman,* Boll. Unione Mat. Ital, (7), 3A (1989), pp. 243-250.

2. T. S. ANGELL, R. E. KLEINMAN, G. F. ROACH, *An inverse transmission problem for the Helmholtz equation,* Inverse Problems, 3 (1987), pp. 149-180.

3. H. BELLOUT, A. FRIEDMAN, *Identification problems in potential theory,* Arch. Rat. Mech.Analysis ,101(1), (1988), pp. 143-160.

4. D. COLTON, R. KRESS , *Integral equations methods in scattering theory,* Wiley, New York, 1983.

5. A. FRIEDMAN , B. GUSTAFSSON, *Identification of the conductivity coefficient in an elliptic equation,* SIAM J. Math. Anal., 18 (1987), pp. 777-788.

6. A. FRIEDMAN, V. ISAKOV, *On the uniqueness in the inverse conductivity problem with one measurement,* Indiana Univ. Math. J., 38 (1989), pp. 553-580.

7. D. ISAACSON, *Distinguishability of conductivities by electric current computed tomography,* IEEE Trans. Medical Imag., MI-5, #2 (1986), pp. 91-95.

8. V. ISAKOV, *On uniqueness of recovery of a discontinuous conductivity coefficient,* Comm.Pure Appl. Math., 41 (1988), pp. 865-877.

9. V. ISAKOV, *Completeness of products and inverse problems for PDE,* J. Diff. Equa., 92(1991), pp. 305-315.

10. V. ISAKOV, *Inverse Source Problems,* AMS Math. Monographs and Surveys, vol. 34,Providence (1990).

11. V. ISAKOV, J. POWELL, *On the inverse conductivity problem with one measurement,*Inverse Problems, 5 (1990), pp. 311-318.

12. V. ISAKOV, *Some inverse problems for elliptic and parabolic equations,* 'Inverse problems for partial differential equations,' SIAM Proc. Series, 1990, pp. 203-214.

13. V. ISAKOV, *On uniqueness in the inverse transmission scattering problem,* Comm. in Part.Diff. Equa., 15 (1990), pp. 1565-1587.

14. R. KOHN, M. VOGELIUS , *Determining conductivity by boundary measurements,* II,Interior results, Comm. Pure Appl. Math., 38 (1985), pp. 643-667.

15. R. KOHN, M. VOGELIUS , *Relaxation of a variational method for impedance computed tomography,* Comm. Pure Appl. Math., 40 (1987), pp. 745-777.

16. P. LAX, R. PHILLIPS , *Scattering theory,* Academic Press, 1967.

17. P.LAX, R.PHILLIPS , *Scattering of sound waves by an obstacle,* Comm. Pure Appl.Math., 30 (1977), pp. 195-233.

18. A. MAJDA ,M. TAYLOR, *Inverse scattering problems for transparent obstacles,electromagnetic waves and hyperbolic systems,* Comm. in Part. Diff. Equa., 2 (1977), pp. 395-438.

19. A. MAJDA, *A representation formula for the scattering operator and the inverse problems for arbitrary bodies,* Comm. Pure Appl. Math., 30 (1977), pp. 169-194.

20. A. NACHMAN, *Reconstructions from boundary measurements,* Ann. of Math., 128 (1988), pp. 531-576.

21. J. SYLVESTER, G. UHLMANN, *A global uniqueness theorem for an inverse boundary value problem,* Ann. of Math., 125 (1987), pp. 153-169

22. J. SYLVESTER, G. UHLMANN, *The Dirichlet-to-Neumann map,* in Inverse Problems in PDE, SIAM, Philadelphia, 1990.

CHAPTER 5

Approximation of the Surface Impedance for a Stratified Medium

Patrick Joly*
Jean E. Roberts*

Abstract:

We consider the problem of calculating the pressure in a fluid domain overlying a stratified elastic medium. The idea is that for such calculations it suffices to know the transmission operator or surface impedance on the boundary between the fluid and elastic mediums instead of the parameters(density and Lamé coefficients) of the elastic medium We define a class of approximate impedance operators in which one could try to identify the best approximate to the true impedance of the elastic medium in the sense that the corresponding pressure is the closest in terms of least squares to that associated with the real medium. Here we treat the direct problem associated with this inverse problem, not the inverse problem itself. We study the stability of the associated initial boundary value problem and its numerical resolution. Two numerical schemes are described, and a stability result is demonstrated. Numerical results are given. For the particular case of one single acoustic layer the approximate impedance corresponds to an absorbing boundary condition, so we obtain as special cases discretizations of absorbing boundary conditions of arbitrary orders.

1 - Introduction

In certain physical problems one is interested in solving the wave equation in some part of the ocean with the objective of identifying a foreign body. However one must first identify the parameters characterizing the earth underneath the ocean floor, i.e. the density and the Lame coefficients. The idea that motivates this study is that it might be interesting to try to find instead of these usual parameters, the impedance of the ocean floor, that is the operator Z which associates to the time derivative of the fluid pressure at the ocean floor its outward normal derivative. However as one should suspect, Z is rarely if ever a differential operator.

---

*INRIA B.P. 105, 78153 Le Chesnay, France.

Thus we are led to define a class of approximate impedance operators which are differential operators and which lead to well posed problems when used to define an initial boundary value problem. It is in this class of operators that one would like to identify the example for which the solution to the associated initial boundary value problem best approximates an observed solution. Here we shall not treat in any detail this identification or inverse problem but shall be concerned with the direct problem. We define a class of approximate impedances and study the corresponding initial boundary value problem, its well posedness and numerical resolution.

In the remainder of section 1 we give a mathematical description of the physical problem. In section 2 we define the impedance operator and in section 3 we give a class of approximate impedance operators. Section 4 is concerned with the stability of the continuous direct problem associated with an approximate impedance operator, and section 5 with its numerical approximation. Two numerical schemes are described and a numerical stability result is demonstrated. Numerical results are presented in section 6. A conclusion and a brief sketch of the inverse problem make up section 7.

Consider a fluid medium Ω_F overlying a stratified elastic medium Ω_S with interface Γ. Ω_S will be identified with the half plane $\{(x,z) \in \mathbb{R}^2 ; z > 0\}$ and Ω_F with the horizontal strip $\{(x,z) \in \mathbb{R}^2 ; -\ell < z < 0\}$. Γ is then the x-axis and the boundary $\Gamma_F = \{(x,z) \in \mathbb{R}^2 ; z = -\ell\}$ is a free boundary.

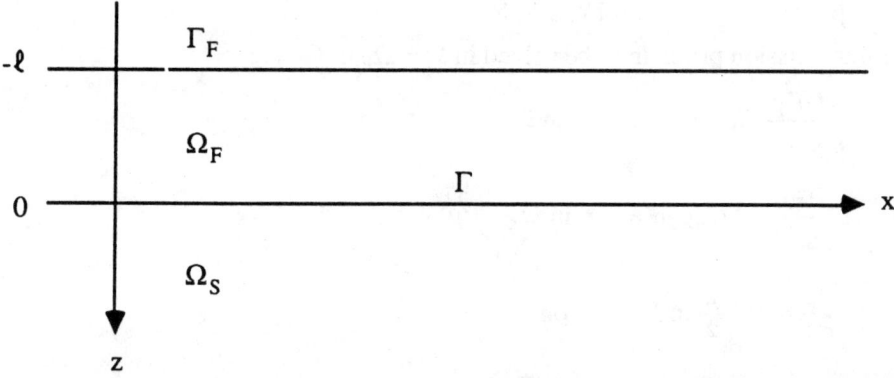

Behavior in Ω_F is governed by the acoustic wave equation :

$$(1.1) \qquad \rho_F \frac{\partial^2 p}{\partial t^2} - \lambda_F \Delta p = f, \qquad \text{in } \Omega_F$$

where p denotes the fluid pressure, ρ_F the density, λ_F the Lame parameter, and f a source term with compact support in Ω_F. The parameters ρ_F and λ_F are taken to be constant and the source term f depends on x and t. In Ω_S the controling equation is the equation of linear elastodynamics :

(1.2) $\rho \dfrac{\partial^2 u}{\partial t^2} - A(\lambda,\mu)u = 0,$ in Ω_S.

The operator $A(\lambda,\mu)$ is just the divergence of the stress tensor $\sigma(u)$:

(1.3) $A(\lambda,\mu) = \text{div } \sigma(u)$

where the stress tensor σ is given by :

(1.4) $\sigma(u) = \begin{pmatrix} (\lambda+2\mu)\dfrac{\partial u_x}{\partial x} + \lambda \dfrac{\partial u_z}{\partial z} & \mu\left(\dfrac{\partial u_x}{\partial z} + \dfrac{\partial u_z}{\partial x}\right) \\ \mu\left(\dfrac{\partial u_x}{\partial z} + \dfrac{\partial u_z}{\partial x}\right) & \lambda \dfrac{\partial u_x}{\partial x} + (\lambda+2\mu)\dfrac{\partial u_z}{\partial z} \end{pmatrix}$

with u he displacement, ρ the density, and λ and μ the Lame coefficients. The parameters ρ, λ and μ are assumed to be functions of z alone ; Ω_S is stratified. The transmission conditions on Γ are the continuity of the normal component of displacement ; which leads to :

(1.5) $-\dfrac{\partial p}{\partial n} = \rho \dfrac{\partial^2 u}{\partial t^2}.n$ on Γ

and the continuity of the stress times n :

(1.6) $\sigma(u)n = -pn$ on Γ

where n is a unit vector normal to Γ.

The boundary condition on the free boundary Γ_F is :

(1.7) $p = 0$ on Γ_F.

Thus the transmission problem to be solved in $\Omega = \Omega_F \cup \Omega_S$ is :

$\rho_F \dfrac{\partial^2 p}{\partial t^2} - \lambda_F \Delta_p = f$ in Ω_F

$\rho \dfrac{\partial^2 u}{\partial t^2} - A(\lambda,\mu)u = 0$ in Ω_S

(1.8) $-\dfrac{\partial p}{\partial n} = \rho \dfrac{\partial^2 u}{\partial t^2}.n$ on Γ

$-\sigma(u)n = pn$ on Γ

$p = 0$ on Γ_F

initial conditions = 0   on $\Omega_F \cup \Omega_S$.

2 - The impedance operator Z

The impedance operator Z associates to the time derivative of the fluid pressure on Γ, the negative of its outward normal derivative there, so that once Z is known, the pressure in Ω_F can be determined without further regard to the underlying elastic medium Ω_S. To define Z, let $\varphi(x)$ be a function defined on Γ and denote by $u^\varphi(x,z,t)$ the solution u of the following problem defined on Ω_S :

$$\rho \frac{\partial^2 u}{\partial t^2} - \text{div}(\lambda(u)) = 0 \qquad \text{in } \Omega_S$$

(2.1) $$-\frac{\partial \sigma(u)n}{\partial t} = \varphi n \qquad \text{on } \Gamma$$

$$\text{zero initial conditions} \qquad \text{on } \Omega_S.$$

Then put :

$$Z(\varphi)(x,t) = \rho \frac{\partial^2 u^\varphi}{\partial t^2} \cdot n \, (x,0,t),$$

so that Z is the composite operator :

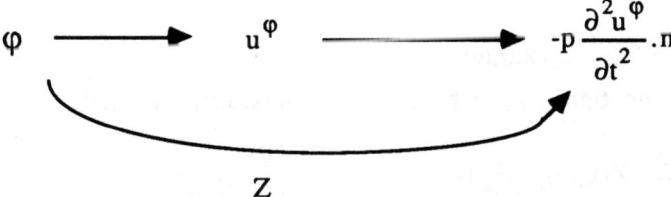

Now the solution p of (1.8) in Ω_F coincides with the solution of the following initial boundary value problem in Ω_F.

$$\frac{\partial^2 p}{\partial t^2} - \Delta p = f \qquad \text{in } \Omega_F$$

$$\frac{\partial p}{\partial n} + Z\left(\frac{\partial p}{\partial t}\right) = 0 \qquad \text{on } \Gamma$$

(2.2)
$$p = 0 \qquad \text{on } \Gamma_F$$

$$\text{zero initial conditions} \qquad \text{in } \Omega_F.$$

Using the Fourier transform in the tangential variable x and the time variable t,

$$\varphi(x,t) \xrightarrow{\mathcal{F}} \hat\varphi(k,\omega),$$

k representing the wave number in the direction x and ω the pulsation, we define the comple impedance $\hat Z(k,\omega)$ of the stratified medium Ω_S so that

$$\widehat{Z\varphi}(k,\omega) = \hat Z(k,\omega)\hat\varphi(k,\omega)$$

as follows :

Let $\hat u(k,z,\omega)$ be the solution $\hat u$ of the following system of ordinary differential equations

$$- \rho\omega^2 \widehat{u} - \widehat{\mathrm{div}\sigma(u)} = 0 \quad \text{for } z > 0$$

(2.3)

$$- i\omega\widehat{\sigma(u)n} = n \qquad \text{at } z = 0,$$

or more explicitly

$$- \frac{d}{dz}\left(\mu(z)\frac{d\widehat{u_x}}{dz}\right) - ik\left(\lambda\frac{d\widehat{u_z}}{dz} + \frac{d}{dz}\,\mu\widehat{u_z}\right) + \left[(\lambda+2\mu)k^2 - \rho\omega^2\right]\widehat{u_x} = 0 \qquad \text{for } z > 0$$

(2.4)

$$- \frac{d}{dz}\left((\lambda+2\mu)\frac{\partial\widehat{u_z}}{\partial z}\right) - ik\left(\mu\frac{d\widehat{u_x}}{dz} + \frac{d}{dz}\lambda\widehat{u_x}\right) + [\mu k^2 - \rho\omega^2]\widehat{u_z} = 0 \qquad \text{for } z > 0$$

$$\mu\frac{d\widehat{u_x}}{dz} + ik\widehat{u_z} = 0 \qquad \text{at } z = 0$$

$$- i\omega\left((\lambda+2\mu)\frac{d\widehat{u_z}}{dz} + ik\lambda\widehat{u_x}\right) = 1 \qquad \text{at } z = 0$$

where u_x and u_z are the x and z components of u. Then the complex impedance $\hat{Z}(k,\omega)$ is given by

(2.5) $\hat{Z}(k,\omega) = - \rho\omega^2\hat{u}_z(k,0,\omega).$

If we introduce the function $K(x,t)$ whose Fourier transform is $\hat{Z}(k,\omega)$,

$$K(x,t) \xrightarrow{\mathfrak{F}} \hat{Z}(k,\omega),$$

the impedance operatoir is non other than the convolution operator

$$(Z\varphi)(x,t) = \int_0^{+\infty}\int_{-\infty}^{+\infty} K(x-\xi,t-\tau)\varphi(\xi,\tau)\,d\xi\,d\tau$$

with kernel $K(x,t)$.

Examples

In the simple case for which the stratified medium Ω_S is in fact homogeneous we can calculate directly the impedance operator Z or more precisely the complex impedance $\hat{Z}(k,\omega)$ of Ω_S. Even though in the model in which we are interested Ω_S is an elastic medium we think it instructive to consider an example where Ω_S is another acoustic medium.

Ω_S a homogeneous acoustic medium

The parameters characterising Ω_F are ρ_F and λ_F and the pressure is denoted by p while the parameters characterizing Ω_S are ρ_S and λ_S and the pressure is p_s. We assume Ω_S and Ω_F are homogeneous, i.e. $\rho_F, \lambda_F, \rho_S$ and λ_S are constant, and we denote the velocities in Ω_F and Ω_S by c_F and c_S respectively, $c^2_F = \lambda_F/\rho_F$ and $c^2_S = \lambda_S/\rho_S$.

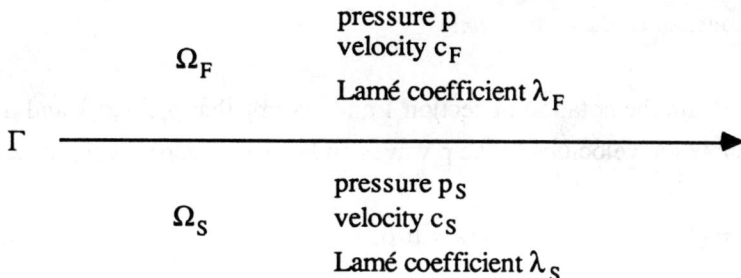

In this case the transmission conditions on Γ are

(2.6) $p = p_S$ on Γ

(2.7) $\lambda_F \dfrac{\partial p}{\partial z} = \lambda_S \dfrac{\partial p_S}{\partial z}$ on Γ,

and the impedance operator Z is given by

(2.8) $Z(\varphi)(x,t) = - \dfrac{\lambda_S}{\lambda_F} \dfrac{\partial p_S}{\partial z}(x,0,t),$

where p_S is the solution in Ω_S of

$$\dfrac{\partial^2 p_S}{\partial t^2} - c_S^2 \Delta p_S = 0 \qquad \text{in } \Omega_S$$

(2.9) $\dfrac{\partial p_S}{\partial t} = \varphi$ on Γ

           zero initial conditions   in Ω_S.

The complex impedance $\hat{Z}(k,\omega)$ is then given by

$$\hat{Z}(k,\omega) = - \dfrac{\lambda_S}{\lambda_F} \dfrac{\widehat{\partial p_S}}{\partial z}(k,0,\omega)$$

with p_S the solution of the ordinary differential equation

$$\dfrac{d^2 \hat{p}_S}{dz^2} + \dfrac{\omega^2}{c_S^2}\left(1 - \dfrac{c_S^2 k^2}{\omega^2}\right)\hat{p}_S = 0 \quad z > 0$$

(2.10)

           $i\omega \hat{p}_S(k,0,\omega) = 1$ $z = 0$.

Solving (2.10) we obtain

$$\hat{p}_S(k,z,\omega) = \hat{p}_S(k,0,\omega)\, e^{-\frac{i\omega}{c_S}\sqrt{1 - \left(\frac{c_S k}{\omega}\right)^2}\, z}$$

and :

(2.11) $\hat{Z}(k,\omega) = \dfrac{1}{c_S}\dfrac{\lambda_S}{\lambda_F}\sqrt{1 - \left(\dfrac{c_S k}{\omega}\right)^2}.$

Ω_S a homogeneous elastic medium

We retain the notation of section 1 and assume that $\rho_F, \lambda_F, \rho, \lambda$ and μ are constant. Let v_p and v_S denote the velocities of the p waves and s waves respectively in Ω_S,

$$v^2_p = (\lambda + 2\mu)/\rho \qquad v^2_S = \mu/\rho.$$

$$\Omega_F \qquad \begin{array}{l} \text{pressure } p_S \\ \text{density } \rho_F \\ \text{velocity } c \\ \text{Lamé coefficient } \lambda_F \end{array}$$

Γ ————————————————————————→

$$\Omega_S \qquad \begin{array}{l} \text{displacement } u \\ \text{density } \rho \\ \text{velocity of p waves } v_p \\ \text{velocity of s waves } v_s \\ \text{Lamé coefficients } \lambda, \mu \end{array}$$

With Ω_S homogeneous, on multiplying by ρ, the system of ordinary differential equations (2.4) becomes

$$v_s^2 \frac{\widehat{\partial^2 u_x}}{\partial z^2} + ik(v_p^2 - v_s^2) \frac{\widehat{\partial u_z}}{\partial z} - (k^2 v_p^2 - \omega^2) \hat{u}_x = 0 \qquad z > 0,$$

$$v_p^2 \frac{\widehat{\partial^2 u_z}}{\partial z^2} + ik(v_p^2 - v_s^2) \frac{\widehat{\partial u_x}}{\partial z} - (k^2 v_s^2 - \omega^2) \hat{u}_z = 0 \qquad z > 0,$$

$$v_s^2 \left(\frac{\widehat{\partial u_x}}{\partial z} + ik\hat{u}_z \right) = 0 \qquad\qquad z = 0,$$

$$-\rho i\omega \left(v_p^2 \frac{\widehat{\partial u_x}}{\partial z} + ik(v_p^2 - 2v_s^2) \hat{u}_z \right) = 1 \qquad z = 0.$$

The solution is thus of the form

$$\hat{u}_x = A(k,\omega)(-k) e^{-iy_p z} + B(k,\omega) y_s e^{-iy_s z}$$
$$\hat{u}_z = A(k,\omega) y_p e^{-iy_p z} + B(k,\omega) k e^{-iy_s z},$$

with :

$$(2.12) \qquad y_s(k,\omega) = \frac{\omega}{v_s} \sqrt{1 - \left(\frac{kv_s}{\omega}\right)^2} \quad \text{and} \quad y_p(k,\omega) = \frac{\omega}{v_p} \sqrt{1 - \left(\frac{kv_p}{\omega}\right)^2},$$

and the constants $A(k,\omega)$ and $B(k,\omega)$ are calculated using the equations on z=0 :

$$A(k,\omega) = \frac{(\omega^2 - 2k^2 v_s^2)}{\Delta \omega \rho} \quad , \quad B(k,\omega) = \frac{2kv_s^2 y_p}{\Delta \omega \rho}$$

with

(2.13) $\Delta(k,\omega) = -\left[(\omega^2 - 2v_s^2 k^2)^2 + 4k^2 v_s^4 y_s y_p \right].$

Hence we have

$$\hat{u}_z(k,0,\omega) = \frac{\omega y_p(k,\omega)}{\rho \Delta(k,\omega)},$$

and (2.5) becomes

(2.14) $\hat{Z}(k,\omega) = \frac{\omega^3 y_p(k,\omega)}{\Delta(k,\omega)}.$

Some properties of the operator Z

We state here without proof some of the properties of the operator Z which we shall try to preserve when we define a class of approximate impedance operators.

(1) Z commutes with the translation in space and translation in time operators :

$$Z(\varphi(x+L)t+\tau)) = (Z\varphi)(x+L,t+\tau).$$

In other words Z is a convolution operator.

(2) Z commutes with the symmetry operator in the variable x :

$$Z(\varphi(-x,t)) = (Z\varphi)(-x,t).$$

(3) Z is positive :

$$\int_0^T \int_{-\infty}^\infty (Z\varphi)(x,t)\varphi(x,t)\, dx\, dt \geq 0$$

for all $T > 0$. This condition assures us that the initial boundary value problem (2.2) is well posed. This property may be expressed in terms of the complex impedance $\hat{Z}$ as follows :

$$\mathrm{Re}(\hat{Z}(k,\omega)) \geq 0 \qquad\qquad \text{for all } (k,\omega) \in \mathbb{R}^2.$$

(4) Z is causal :

$$\varphi(x,t) = 0 \text{ for all } t \leq T \quad\Rightarrow\quad (Z\varphi)(x,T) = 0.$$

In terms of $\hat{Z}$ this property is that the function $\omega \to \hat{Z}(k,\omega)$ may be extended analytically to the complex half plane $\mathrm{Im}\,\omega > 0$.

(5) Z is real :

$$\varphi(x,t) \in \mathbb{R} \text{ for all } (x,t) \quad\Rightarrow\quad (Z\varphi)(x,t) \in \mathbb{R} \text{ for all } (x,t).$$

The translation of this property into terms of $\hat{Z}$ is :

$$\hat{Z}(-k,\omega) = \hat{Z}(k,-\omega) = \overline{\hat{Z}(k,\omega)}.$$

We shall see in the next section that we can easily preserve all of the above properties for the class of approximate impedances that we shall define with the exception of (3), positivity. This property shall be replaced by the weaker property that the initial boundary value problem (2.2) corresponding to an approximate impedance operator be well posed.

3 - A class of approximate impedance operators

We denote by Σ the class of impedance operators Z corresponding to a stratified medium Ω_S and by $\hat{\Sigma}$ the class of the associated impedances $\hat{Z}$. Given that the ultimate objective is to solve an identification problem in a class Σ_a of approximate impedance operators, this class Σ_a should be constructed in such a way as to satisfy the following criteria :

1) Each element of Σ_a should be determined by a "small" number of parameters.
2) The class Σ_a should be large enough to "well approximate" all elements of Σ.
3) Each element of Σ_a should lead to a well posed initial boundary value problem.
4) The solution of the initial boundary value problem corresponding to an element of Σ_a should be "simple".

Evidently criteria 1) and 2) are at odds. From a practical point of view the reduction of the number of parameters is necessary in order to have a plausible numerical approach to the identification problem. Thus we have favored 1) over 2). Criterion 3), which obviously can not be violated, also goes against criterion 2) in that it restricts the set Σ_a. We shall see that it is indeed this criterion that limits us in our attempt to satisfy 2). Finally criterion 4) is equally necessary for the feasibility of our program in that the use of an optimization algorithm requires a succession of resolutions of the direct problem.

In fact it is criterion 4) that has guided us in our choice of Σ_a. In general the difficulty in working directly with an operator Z in Σ comes from the nonlocal nature in space and in time of these operators, which makes their numerical treatment prohibitif. However it is well known that certain convolution operators are easily calculated by solving a system of differential equations. Such is the case when the kernel of convolution is a sum of exponential functions or equivalently when its fourier transform, here the complex impedance $\hat{Z}(k,\omega)$, is a rational function of (k,ω), whence the idea to choose Σ_a as a subclass of the operators whose complex impedance is a rational function.

This idea has already been exploited by Engquist and Majda [2] and [3] for constructing absorbing boundary conditions. In fact, the problem of absorbing boundary conditions appears as a special case of the problem treated here in which the medium Ω_S is identical in nautre to the medium Ω_F. The impedance complex given by (2.11) is :

$$\hat{Z}(k,\omega) = \frac{1}{c} \sqrt{1 - \left(\frac{ck}{\omega}\right)^2} \, ,$$

$c = c_S = c_F$. In this case a method that has proven quite effective is to use the Pade approximations of the function $\sqrt{1 - x^2}$ around $x = 0$. These approximations are given by

(3.1) $$\sqrt{1-x^2} \simeq 1 - \sum_{p=1}^{N} \frac{a_p x^2}{\alpha_p x^2 + 1}$$

where for the approximations of odd order, $2N+1$, the coefficients are given by

$$a_p = \frac{2}{2N+1} \sin^2 \frac{p\pi}{2N+1}$$

(3.2)

$$\alpha_p = -\cos^2 \frac{p\pi}{2N+1} \, ,$$

and for the approximations of even order, $2N$, the coefficients are given by

$$a_p = \frac{1}{N} \sin^2 \frac{p\pi}{2N} , \quad p = 1,...,N-1 \; ; \; a_N = \frac{1}{2N}$$

(3.3)

$$\alpha_p = - \cos^2 \frac{p\pi}{2N} , \quad p = 1,...,N-1 \; ; \; \alpha_N = 0.$$

It is natural to demand that the class Σ_a contain all operators Z whose corresponding impedance complex $\hat{Z}$ is of the form (3.1), and we have chosen to take Σ_a to be contained in the class of operators whose complex impedance is of the following form

(3.4) $$\hat{Z}(k,\omega) = \frac{1}{c} \left(1 - \sum_{p=1}^{N} \frac{a_p k^2 - b_p \omega^2 + c_p i\omega + d_p}{\alpha_p k^2 - \beta_p \omega^2 + \gamma_p i\omega + \delta_p} \right),$$

with real coefficients $c, a_p, b_p, c_p, d_p, \alpha_p, \beta_p, \gamma_p, \delta_p, \; p=1,...,N$.

Several comments on the form (3.4) are in order. By demanding that the coefficients be real we respect the reality, property (5), of the true impedance. By construction we have respected property (1), invariance under translation, and property (4), causality. The spatial symmetry, property (2), is guaranteed by the fact that the rational function $\hat{Z}(k,\omega)$ is an even function of k. We have required that the "integer part" of the rational function be constant. This is practically required in order to obtain a well posed problem. We shall see in the next section what other requirements must be imposed on the coefficients to obtain a well posed problem. Finally the choice of the form (3.4), with $\hat{Z}(k,\omega)$ given as a sum of simple elements, is justified by the expression of $Z(\varphi)$ in the physical variables (x,t) which makes the numerical resolution of the problem quite simple as we shall see in section 5.

We shall define Σ_a to be the class of operators Z such that :

(3.5) $\hat{Z}(k,\omega)$ is of the form (3.4)

(3.6)   the associated inital boundary value problem (2.2) is well posed.

We denote by $\hat{\Sigma}_a$ the class of corresponding complex impedances.

4 - The direct problem

Given an $8N+1$ tuple $m \in \mathbb{R}^{8N+1}$,

$m = (c, (a_p, b_p, c_p, d_p, \alpha_p, \beta_p, \gamma_p, \delta_p), p=1,...,N)$ we denote by $\hat{Z}_m$ the comple impedance of the form (3.4) corresponding to m and by Σ_m the impedance operator. The problem direct associated to m is the problem (2.2) for $Z=Z_m$:

$$\frac{\partial^2 p}{\partial t^2} - \Delta p = f \qquad\qquad \text{in } \Omega_F$$

$$\frac{\partial p}{\partial n} + Z_m\left(\frac{\partial p}{\partial t}\right) = 0 \qquad\qquad \text{on } \Gamma$$

$$p = 0 \qquad\qquad \text{on } \Gamma_F$$

$$p = \frac{\partial p}{\partial t} = 0 \qquad\qquad \text{in } \Omega_F \text{ at } t = 0.$$

For simplicity we have assumed $c_F = \lambda_F = \rho_F = 1$. The first question that arises in the study of such a problem is whether the problem is well posed. Does there exist a unique solution depending continuously on f. The answer of course depends on m and will give us an explicit criterion for determining which parameters $m \in \mathbb{R}^{8N+1}$ correspond to an element of Σ_a.

Given an $m \in \mathbb{R}^{8N+1}$, $\hat{Z}_m(k,\omega)$ is a rational function of (k,ω) and may be written in the form

$$(4.1) \qquad \hat{Z}_m(k,\omega) = \frac{Q(ik,i\omega)}{R(ik,i\omega)},$$

with

$$Q(ik,i\omega) = \prod_{p=1}^{N} (\alpha_p k^2 - \beta_p \omega^2 + \gamma_p i\omega + \delta_p)$$

$$+ \sum_{p=1}^{N} \left((a_p k^2 - b_p \omega^2 + c_p i\omega + d_p) \prod_{\substack{q=1 \\ q \neq p}}^{N} (\alpha_q k^2 - \beta_q \omega^2 + \gamma_q i\omega + \delta_q) \right)$$

$$R(ik,i\omega) = c \prod_{p=1}^{N} (\alpha_p k^2 - \beta_p \omega^2 + \gamma_p i\omega + \delta_p).$$

The polynomials Q and R have real coefficients and are of degree less than or equal to 2N, and we may write

$$R\left(ik,i\omega\right) \hat{Z}_m\left(k,\omega\right) \hat{\varphi} = Q(ik,i\omega)\hat{\varphi}.$$

Applying the inverse Fourier transform we obtain

$$R\left(\frac{\partial}{\partial x},\frac{\partial}{\partial t}\right) Z_m\varphi = Q\left(\frac{\partial}{\partial x},\frac{\partial}{\partial t}\right)\varphi$$

which leads to the following boundary condition

$$R\left(\frac{\partial}{\partial x},\frac{\partial}{\partial t}\right)\frac{\partial p}{\partial n} + Q\left(\frac{\partial}{\partial x},\frac{\partial}{\partial t}\right)\frac{\partial p}{\partial t} = 0 \quad \text{on } \Gamma.$$

Indeed we impose that R and Q are of order 2N by assuming $\beta_p \neq 0$ and $a_p \neq 0$, p=1,...,N. We also assume that

$$\frac{\alpha_p}{\beta_p} \neq \frac{\alpha_q}{\beta_q} \quad \text{if } p \neq q.$$

Since the stability of the problem depends only on the terms of highest degree we introduce the homogeneous polynomial $\tilde{R}(ik,i\omega)$ and $\tilde{Q}(ik,i\omega)$ consisting of the terms of $R(ik,i\omega)$ and $Q(ik,i\omega)$ respectively of maximal degree :

$$\tilde{Q}(ik,i\omega) = \prod_{p=1}^{N}(\alpha_p k^2 - \beta_p \omega^2) + \sum_{p=1}^{N}\left((a_p k^2 - b_p \omega^2)\prod_{\substack{q=1 \\ q \neq p}}^{N}(\alpha_q k^2 - \beta_q \omega^2)\right)$$

$$\tilde{R}(ik,i\omega) = c \prod_{p=1}^{N}(\alpha_p k^2 - \beta_p \omega^2),$$

and consider the condition

$$\tilde{R}\left(\frac{\partial}{\partial x}, \frac{\partial}{\partial t}\right)\frac{\partial p}{\partial n} + \tilde{Q}\left(\frac{\partial}{\partial x}, \frac{\partial}{\partial t}\right)\frac{\partial p}{\partial t} = 0 \quad \text{on } \Gamma$$

and the problem

$$\frac{\partial^2 p}{\partial t^2} - \Delta p = f \qquad \text{in } \Omega = \{(x,z): z < 0\}$$

(4.3)
$$\tilde{R}\left(\frac{\partial}{\partial x}, \frac{\partial}{\partial t}\right)\frac{\partial p}{\partial n} + \tilde{Q}\left(\frac{\partial}{\partial x}, \frac{\partial}{\partial t}\right)\frac{\partial p}{\partial t} = 0 \quad \text{on } \Gamma$$

$$p = \frac{\partial p}{\partial t} = 0 \qquad \text{in } \Omega \text{ at } t = 0,$$

where for simplicity we have replaced the domain Ω_F by the half space Ω.

The stability of the system (4.3) may be studied by the method of normal modes ; see [4], [5], and [6] for a detailed development of this theory. A normal mode is a solution of the equations

(4.4)
$$\frac{\partial^2 p}{\partial t^2} - \Delta p = 0 \quad \text{in } \Omega$$

(4.5)
$$\tilde{R}\left(\frac{\partial}{\partial x}, \frac{\partial}{\partial t}\right)\frac{\partial p}{\partial n} + \tilde{Q}\left(\frac{\partial}{\partial x}, \frac{\partial}{\partial t}\right)\frac{\partial p}{\partial t} = 0 \quad \text{on } \Gamma$$

of the form

$$p = e^{(h+ik)x + (\ell+i\lambda)z + (\eta+i\omega)t} = e^{hx+\ell y+\eta t}e^{i(kx+\lambda z+\omega t)}$$

with amplitude $e^{hx+\ell y+\eta t}$, frequency ω, and velocity vector $v = (v_x, v_z) = (-k/\omega, -\lambda/\omega)$. Since p satisfies (4.4) we have the dispersion relation

$$(\eta+i\omega)^2 - (h+ik)^2 - (\ell+i\lambda)^2 = 0,$$

and since for t fixed, p should remain bounded in Ω we impose $h=0$ and $\ell \geq 0$. Thus we have :

(4.6)
$$\ell+i\lambda = \sqrt{(\eta+i\omega)^2 + k^2}.$$

As p is also a solution of (4.5) we also have

(4.7)
$$\tilde{R}(ik,\eta+i\omega)(\ell+i\lambda) + \tilde{Q}(ik,\eta+i\omega)(\eta+i\omega) = 0.$$

Combining (4.6) with (4.7) we obtain the so called *characteristic equation* of problem (4.3) :

(4.8)
$$F(k,\eta,\omega) = \sqrt{k^2 + (\eta+i\omega)^2}\,\tilde{R}(ik,\eta+i\omega) + (\eta+i\omega)\tilde{Q}(ik,\eta+i\omega) = 0.$$

It is in terms of the zeros of this equation that we can describe the well posedness of problem (4.3).

Problem (4.3) and hence (4.1) is *strongly well posed* if there is no solution of (4.8) with $\eta \geq 0$. It is *weakly well posed* if there is no solution with $\eta > 0$. It is *strongly ill posed* if there exists a solution with $\eta > 0$. For a strongly well posed problem the energy in Ω at any time $T > 0$ can be bounded in the L_2-norm of the source term f. For a weakly well posed problem the energy can be bounded in terms of a higher order Sobolev norm of f, the order needed depending on the multiplicity of the root with $\eta = 0$. One says in this case that derivatives are lost. Finally, for an ill posed problem the energy in Ω grows in time at a completely uncontroled rate. For precise statements of these estimates we refer the reader to [5] or [6].

Before translating these conditions into terms of the parameters m we make a heuristic comment on the numerical approximation of problem (4.1), or (4.3). Denote by Σ^S_a the set of operators Z for which $\hat{Z}$ is of the form (3.4) and for which the problem (4.3) is strongly well posed and by Σ^W_a the set of operators Z for which $\hat{Z}$ is of the form (3.4) and for which the problem (4.3) is weakly well posed. The set Σ^S_a of operators having the zeros of its characteristic equation (4.8) in the open set $\eta < 0$ is open. Σ^W_a is the boundary between Σ^S_a and the set of operators giving rise to an ill posed problem. This indicates that the numerical approximation of a weakly well posed problem is more delicate than that of a strongly well posed problem. A "good" approximation of a strongly well posed problem should be stable, where as a weakly well posed problem may be approached by a numerical problem that is stable or by a numerical problem that is unstable.

To describe the subset of $\mathbb{R}^{8N+1}$ of parameters which correspond to an operator giving rise to a strongly or weakly well posed problem we use a characterization of well posedness developed in [8]. Since $\tilde{R}$ and $\tilde{Q}$ are homogeneous of the same degree, (4.7) implies that

$$\ell + i\lambda = - \frac{\tilde{Q}\left(\frac{ik}{\eta+i\omega}, 1\right)}{\tilde{R}\left(\frac{ik}{\eta+i\omega}, 1\right)} (\eta+i\omega).$$

Defining

(4.9) $$r(\theta) = \frac{\tilde{Q}(\theta,1)}{\tilde{R}(\theta,1)}, \quad \theta \in \mathbb{C},$$

we may write

$$\ell + i\lambda = - \frac{r(\theta)}{\theta} ik, \text{ with } \theta = \frac{ik}{\eta+i\omega}.$$

We may now characterize the strong or weak well posedness of (4.1) and (4.3) in terms of the poles and zeros of $r(\theta)/\theta$.

Problem (4.3) and hence (4.1) is strongly well posed if the zeros and poles of $r(\theta)/\theta$ are real, simple and alternate along the real axis and $r(\theta) > 0$, for $\theta \in [-1,1]$. It is weakly well posed if the zeros and poles of $r(\theta)/\theta$ are real, simple and alternate along the real axis and $r(0) > 0$. (cf. [8], theorems 1 and 2).

Decomposing $r(\theta)/\theta$ into partial fractions we obtain

$$\frac{r(\theta)}{\theta} = \psi_* \theta + \psi_0 \frac{1}{\theta} + \sum_{\{p; \alpha_p \neq 0\}} \psi_p \left(\frac{1}{\theta-\theta_p} + \frac{1}{\theta+\theta_p} \right)$$

with

$$\psi_* = \frac{1}{c} \sum_{\{p;\alpha_p=0\}} \frac{a_p}{\beta_p}$$

$$\psi_0 = \frac{1}{c} \left\{ 1 - \sum_{p=1}^{N} \frac{a_p}{\beta_p} \right\}$$

$$\psi_p = -\frac{1}{2c} \left(\frac{a_p}{\alpha_p} - \frac{b_p}{\beta_p} \right) \qquad \text{for all } p \text{ with } \alpha_p \neq 0$$

$$\theta_p = \sqrt{\frac{\beta_p}{\alpha_p}} \qquad \text{for all } p \text{ with } \alpha_p \neq 0.$$

In order that the poles of $r(\theta)/\theta$ be real and simple it is necessary and sufficient that β_p and α_p have the same sign. There will be exactly one zero between a pair of consecutive poles if and only if the coefficients ψ_p, $p=0,...,N$ are of the same sign and ψ_* is of the opposite sign. Since the coefficient of $1/\theta$ determines the sign of $r(0)$, the conditions for having a weakly well posed problem are

$$\beta_p/\alpha_p > 0 \quad \text{for all } p \text{ with } \alpha_p \neq 0 \quad p=1,...,N$$

$$\psi_p > 0 \qquad \text{for all } p \text{ with } \alpha_p \neq 0 \quad p=1,...,N$$

(4.10)

$$\psi_0 > 0$$

$$\psi_* \leq 0.$$

To have further that $r(\theta) > 0$ for $-1 \leq \theta \leq 1$ we need in addition to the conditions (4.10) that no pole of $r(\theta)$ lies in $[-1,1]$ and that $r(1) > 0$:

(4.11) $\beta_p/\alpha_p > 1$ \qquad for all p with $\alpha_p \neq 0$; $p=1,...,N$

(4.12) $r(1) = \dfrac{1}{c}\left(1 - \displaystyle\sum_{p=1}^{N} \frac{a_p - b_p}{\alpha_p - \beta_p} \right) > 0.$

Remark 4.1

Suppose that m is such that (4.10) and (4.11) hold but instead of (4.12) we have

(4.13) $r(1) = \dfrac{1}{c}\left(1 - \displaystyle\sum_{p=1}^{N} \frac{a_p - b_p}{\alpha_p - \beta_p} \right) = 0.$

Then problem (4.1, 4.3) is only weakly well posed. However, the solution (k,η,ω) of the characteristic equation (4.8) with $\eta=0$ also has $k=0$. It is said to be a "generalized eigenvalue of the first type". In certain cases it has been shown that one can obtain interior energy estimates for such problems resembling those obtained for strongly well posed problems, cf. [7]. We shall see in our numerical examples of paragraph 6 that such a problem exhibits the behavior of a strongly well posed problem.

We consider the approximation of the true impedance operators Z calculated in paragraph 2, (2.11) and (2.14), by elements Z_m in Σ^S_a and Σ^W_a.

The Pade approximations of (2.11) are given by (3.1) with (3.2) and (3.3) :

$$\widehat{Z}_{2N+1}(k,\omega) = \frac{1}{c}\left\{1 - \sum_{p=1}^{N} \frac{a_p k^2}{\alpha_p k^2 - \beta_p \omega^2}\right\}$$

$$c = c_S \frac{\lambda_F}{\lambda_1}$$

(4.14)

$$a_p = \frac{2}{2N+1} \sin^2 \frac{p\pi}{2N+1} c_S^2 \quad p=1,...,N$$

$$\alpha_p = -\cos^2 \frac{p\pi}{2N+1} c_S^2 \qquad p=1,...,N$$

$$\beta_p = -1 \qquad\qquad\qquad p=1,...,N,$$

and :

$$\widehat{Z}_{2N}(k,\omega) = \frac{1}{c}\left\{1 - \sum_{p=1}^{N} \frac{a_p k^2}{\alpha_p k^2 - \beta_p \omega^2}\right\}$$

$$c = c_S \frac{\lambda_F}{\lambda_1}$$

(4.15)

$$a_p = \frac{1}{N} \sin^2\left(\frac{p\pi}{2N}\right) c_S^2 \quad p=1,...,N-1, \quad a_N = \frac{1}{2N} c_1^2$$

$$\alpha_p = -\cos^2\left(\frac{p\pi}{2N}\right) c_S^2 \quad p=1,...,N-1, \quad \alpha_N = 0$$

$$\beta_p = -1 \qquad\qquad\qquad p=1,...,N.$$

We can verify immediately that Z_N and Z_{2N+1} belong to Σ^W_a, but it is also obvious that condition (4.11) can not be satisfied for large N if $c_S > c$, and Z_{2N} and Z_{2N+1} will not be in Σ^S_a.

We consider the approximations of lowest order

(4.16) $$\widehat{Z}_1(k,\omega) = \frac{\lambda_S}{\lambda_F c_S}$$

(4.17) $$\widehat{Z}_2(k,\omega) = \frac{\lambda_S}{\lambda_F c_S}\left(1 - \frac{\frac{1}{2} c_S^2 k^2}{\omega^2}\right)$$

(4.18) $$\widehat{Z}_3(k,\omega) = \frac{\lambda_S}{\lambda_F c_S}\left(1 - \frac{\frac{1}{2} c_S^2 k^2}{\omega^2 - \frac{1}{4} c_S^2 k^2}\right),$$

and the corresponding functions $r_N(\theta)$

$$r_1(\theta) = [\lambda_S/(\lambda_F c_S)]$$

(4.19) $$r_2(\theta) = [\lambda_S/(\lambda_F c_S)](1 - 1/2\ c_S^2\theta^2)$$

$$r_3(\theta) = [\lambda_S/(\lambda_F c_S)][1 - (1/2\ c_S^2\theta^2)/(1 - 1/4\ c_S^2\theta^2)].$$

Since the constants λ_S λ_F and c_S are always positive $r_1(\theta)$ is always positive and $Z_1 \in \Sigma^S_a$ but $r_2(1)$ is negative once $c_S > \sqrt{2}$ and $r_3(1)$ is negative once $c_S > (2\sqrt{2})/3$.

The function $r(\theta)$ corresponding to the true impedance $\hat{Z}(k,\omega)$ given by (2.11) is :

$$(4.20) \qquad r(\theta) = \frac{\lambda_S}{c_S\lambda_F} \sqrt{1-c_S^2\theta^2} \ ,$$

whose graph is a semi-ellipse with axes of lengths $\lambda_S/(\lambda_F c_S)$ and $1/c_S$. It is clear that if $c_S \gg 1$ we can not approximate $r(\theta)$ well for large values of θ by functions $r_a(\theta)$ which remain positive in the interval $[-1,1]$. We can find approximations $r_a(\theta)$ better that $r_1(\theta)$ which are positive in $[-1,1]$, for example :

$$(4.21) \qquad r_a(\theta) = \frac{\lambda_S}{c_S\lambda_F} (1-(1-\epsilon)\theta),$$

but they will not be more than first order at $\theta=0$ and will not be good approximations for $\theta=1$ or -1, cf.Figure 1. This is why we are interested in the class Σ^W_a as well as the class Σ^S_a.

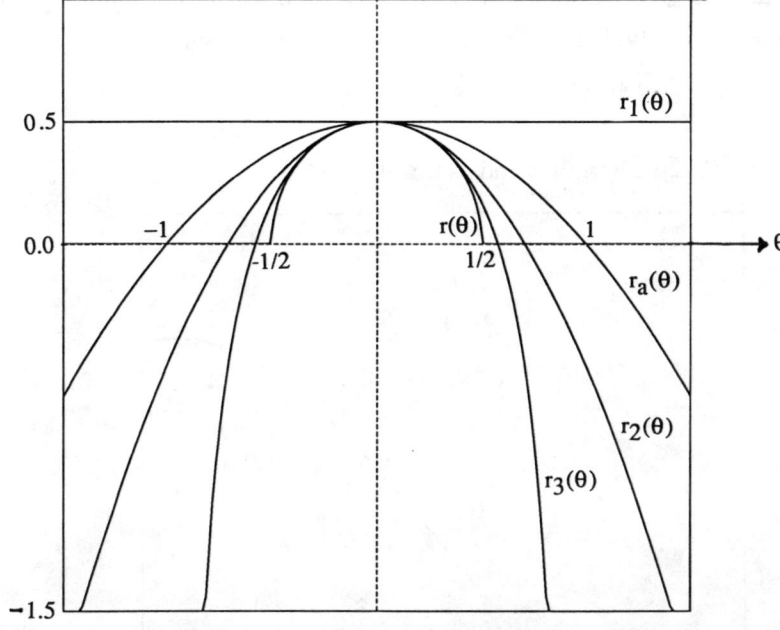

Figure 1. The functions $r(\theta)$, $r_1(\theta)$, $r_2(\theta)$, $r_3(\theta)$, and $r_a(\theta)$.

A general expression for the Pade approximations of (2.14) is not so readily obtained. We give the expressions for the approximations of lowest order

$$\hat{Z}_1(k,\omega) = \frac{1}{v_p}$$

$$(4.22) \qquad \hat{Z}_2(k,\omega) = \frac{1}{v_p}\left(1 - \frac{ak^2}{\omega^2}\right)$$

$$\hat{Z}_3(k,\omega) = \frac{1}{v_p}\left(1 - \frac{ak^2}{\omega^2 - \alpha k^2}\right)$$

and the corresponding functions $r_N(\theta)$

$$r_1(\theta) = \frac{1}{v_p}$$

(4.23) $$r_2(\theta) = \frac{1}{v_p}(1 - a\theta^2)$$

$$r_3(\theta) = \frac{1}{v_p}\left(1 - \frac{a\theta^2}{1 - \alpha\theta^2}\right),$$

where

$$a = \left(\frac{v_p^2}{2} + 4\frac{v_s^3}{v_p} - 4v_s^2\right)$$

$$\alpha = \frac{1}{a}\left(\frac{v_p^4}{8} + 4v_s^4 - 2\frac{v_s^3}{v_p}(v_p^2 + v_s^2)\right) - 4v_s^4\left(\frac{v_s}{v_p} - 1\right).$$

We observe that $Z_1 \in \Sigma^S_a$ for all ρ, λ, μ. Z_2 is in Σ^W_a if $a \geq 0$ and $Z_2 \in \Sigma^S_a$ if $0 < a < 1$. However Z_2 gives a strongly ill posed problem if $a < 0$. We may write

(4.24) $$\frac{a}{v_p^2} = f(v) = \frac{1}{2} - 4v^2(1-v),$$

with $v = v_s/v_p = \sqrt{\mu/\lambda + 2\mu}$. For all λ and μ, $0 \leq v \leq \sqrt{2}/2$.

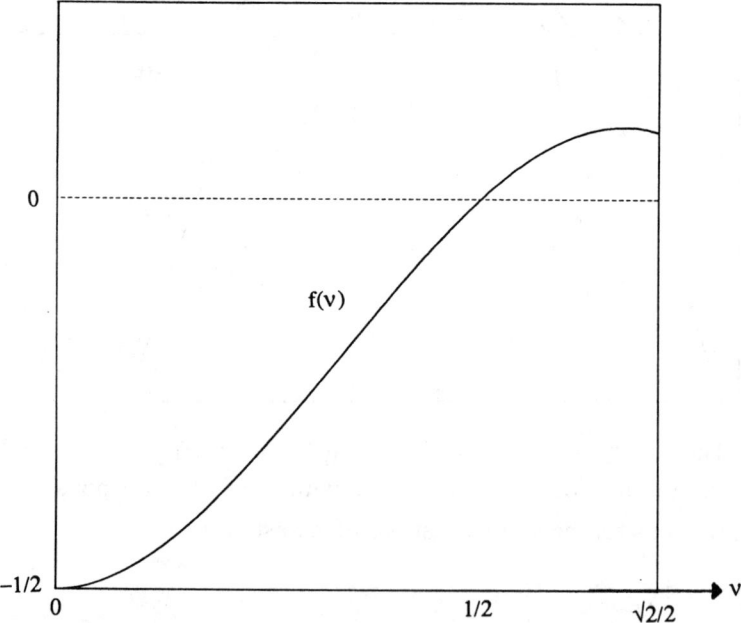

Figure 2. The function $f(v)$.

We see that we have a strongly well posed problem if $0 < v < 1/2$ and $1/2 - 1/v_p^2 < 4v^2(1-v)$ which is the case for instance if $0 < v < 1/2$ and $v_p < \sqrt{2}$. The problem is weakly well posed if $0 < v < 1/2$, but if $1/2 < v < \sqrt{2}/2$ we have an ill posed problem. For v in $[1/2, \sqrt{2}/2]$ the function $r(\theta)$ corresponding to (2.14) is concave upward at $\theta = 0$ and thus can not be approximated to order greater than 1 at $\theta = 0$ even by a function $r(\theta)$ corresponding to a complex impedance $\hat{Z}(k,\omega) \in \hat{\Sigma}^W_a$. This is an example of where condition 3) of paragraph 3 restricts us in our attempt to satisfy condition 2), and this represents a certain limitation to the proposed procedure.

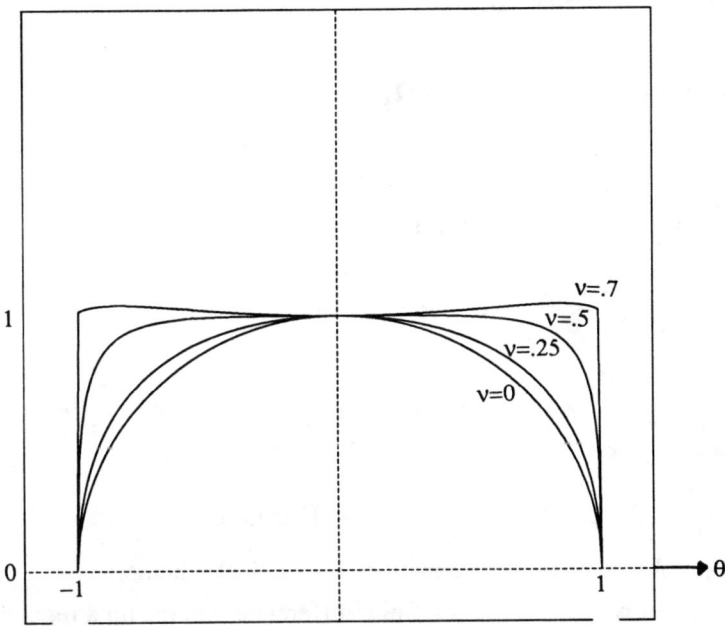

Figure 3. The functions $r(\theta)$ for $v_p=1$ and $v=0$, 1/4, 1/2, and $\sqrt{2}/2$.

5 - Numerical resolution of the direct problem

The form (3.4) for the complex impedance $\hat{Z}$ was chosen to facilitate the resolution of the direct problem (4.1). Indeed

$$(5.1) \qquad \left(Z\widetilde{\frac{\partial p}{\partial \tau}}\right)(k,\omega) = \frac{i\omega}{c}\left\{\hat{p}(k,\omega) - \sum_{p=1}^{N}\left[\frac{a_pk^2-b_p\omega^2+c_pi\omega+d_p}{\alpha_pk^2-\beta_p\omega^2+\gamma_pi\omega+\delta_F}\right]\hat{p}(k,\omega)\right\}$$

is equivalent to the system

$$(5.2) \qquad \left(Z\widetilde{\frac{\partial p}{\partial \tau}}\right)(k,\omega) = \frac{i\omega}{c}\left\{\hat{p}(k,\omega) - \sum_{p=1}^{N}\hat{\varphi}_p(k,\omega)\right\}$$

$$\left[\alpha_pk^2-\beta_p\omega^2+\gamma_pi\omega+\delta_p\right]\hat{\varphi}_p(k,\omega) = \left[a_pk^2-b_p\omega^2+c_pi\omega+d_p\right]\hat{p}(k,\omega),$$

where we have introduced the auxillary variables $\varphi_p(x,t)$ having Fourier transform $\hat{\varphi}_p(k,\omega)$. Applying the inverse Fourier transform to (5.2) we obtain

$$Z\frac{\partial p}{\partial t} = \frac{1}{c}\left\{\frac{\partial p}{\partial t} - \sum_{p=1}^{N}\frac{\partial \varphi_p}{\partial t}\right\}$$

(5.3)

$$\beta_p\frac{\partial^2 \varphi_p}{\partial t^2} - \alpha_p\frac{\partial^2 \varphi_p}{\partial x^2} + \gamma_p\frac{\partial^2 \varphi_p}{\partial t} + \delta_p\varphi_p = b_p\frac{\partial^2 p}{\partial t^2} - a_p\frac{\partial^2 p}{\partial x^2} + c_p\frac{\partial p}{\partial t} + d_p p.$$

Thus problem (4.1) is equivalent to

$$\frac{\partial^2 p}{\partial t^2} - \Delta p = f \qquad \text{in } \Omega_F$$

$$p = 0 \qquad \text{on } \Gamma_F$$

$$p = \frac{\partial p}{\partial t} = 0 \qquad \text{in } \Omega_F \text{ at } t = 0$$

(5.4)

$$\frac{\partial p}{\partial n} + \frac{1}{c}\left(\frac{\partial p}{\partial t} - \sum_{p=1}^{N}\frac{\partial \varphi_p}{\partial t}\right) = 0 \quad \text{on } \Gamma$$

$$\beta_p\frac{\partial^2 \varphi_p}{\partial t^2} - \alpha_p\frac{\partial^2 \varphi_p}{\partial x^2} + \gamma_p\frac{\partial \varphi_p}{\partial t} + \delta_p \varphi_p = b_p\frac{\partial^2 p}{\partial t^2} - a_p\frac{\partial^2 p}{\partial x^2} + c_p\frac{\partial p}{\partial t} + d_p p \qquad \text{on } \Gamma$$

$$\varphi_p = \frac{\partial \varphi_p}{\partial t} = 0, \quad p = 1,...,N, \qquad \text{on } \Gamma \text{ at } t = 0,$$

and it is for this problem that we shall construct a numerical scheme.

The scheme that we have used is a finite difference scheme on a regular grid of squares of side length h. The unknowns are the values p^n_{ij} of p at the points (ih,jh) in $\Omega_F \cup \Gamma$ at instant $n\Delta t$ and the values φ^n_{pi} of φ_p at (ih,Jh) on the boundary Γ, Z=Jh, at instant $n\Delta t$, p=1,...,N.

The scheme for the equation in Ω_F is the standard five point scheme for the Laplacian and the explicit centered difference scheme for the second time derivative :

(5.5) $$\frac{p^{n+1}_{ij} - 2p^n_{ij} + p^{n-1}_{ij}}{\Delta t^2} - \frac{p^n_{ij+1} + p^n_{ij-1} + p^n_{i+1j} + p^n_{i-1j} - 4p^n_{ij}}{h^2} = f^n_{ij}.$$

For the equation on Γ we first considered a centered explicit scheme. The first and second time derivatives are approximated by standard centered differences and the second derivative in x is approximated by the standard centered difference :

(5.6) $$\frac{\partial u}{\partial t}(ih,Jh,n\Delta t) \sim \frac{1}{2\Delta t}\left(u^{n+1}_{iJ} - u^{n-1}_{iJ}\right)$$

(5.7) $$\frac{\partial^2 u}{\partial t^2}(ih,Jh,n\Delta t) \sim \frac{1}{\Delta t^2}\left(u^{n+1}_{iJ} - 2u^n_{iJ} + u^{n-1}_{iJ}\right)$$

(5.8) $$\frac{\partial^2 u}{\partial x^2}(ih,Jh,n\Delta t) \sim \frac{1}{h^2}\left(u^n_{i+1J} - 2u^n_{iJ} + u^n_{i-1J}\right)$$

where u is either p or φ_p. The normal derivative at Γ is approximated with a centered difference introducing fictions values p^n_{iJ+1}

(5.9) $$\frac{\partial p}{\partial z}(ih,Jh,n\Delta t) \sim \frac{1}{2h}\left(p^n_{i,J+1} - p^n_{i,J-1}\right),$$

and these fictions values are then eliminated by supposing that (5.5) holds on Γ, for j=J.

Thus we obtain N+1 equations with which we can calculate explicitly p^{n+1}_{iJ} and φ^{n+1}_{pi}, p=1,...,N, knowing all of the values, $p^n_{i'j}$ and $\varphi^n_{pi'}$, of p and φ_p on Γ and of p in Ω_F at the preceeding time $n\Delta t$.

We note that this scheme is identical to the one we would obtain using P_1 finite elements with mass lumping.

With this scheme we have obtained good numerical results for the case $Z_m \in \Sigma^S_a$, i.e. when the problem is strongly well posed ; see section 6. However in most of the cases for which the problem is only weakly well posed the scheme became unstable, the exception being those corresponding to the case of a generalized eigenvalue of the first kind, of Remark 4.1.

In order to calculate with impedance operators Z_m in Σ^W_a we have devised a second numerical scheme. We keep the scheme (3.5) to calculate the interior values p^{n+1}_{ij}, j < J. In order to conserve the second order precision of the scheme we have kept a centered scheme, but this time it is centered at (ih,(J-1)h). For the first equation on Γ in (5.4), with only first order time derivatives, the approximation is centered in time at instant n+1/2. Thus the time derivative is approximated by the average between (ih,Jh) and (ih,(J-1)h) of the centered difference at instant n+1/2 :

$$(5.10) \qquad \frac{\partial u}{\partial t}(ih,(J-1/2)h,n\Delta t) \sim \frac{1}{2\Delta t}\left(u^{n+1}_i - u^n_i\right),$$

where u^n_i is either $(p^n_{iJ}+p^n_{iJ-1})/2$ or φ^n_{pi}, and the normal derivative is approximated by the average between instants n+1 and n of the centered difference at (ih,(J-1/2)h) :

$$(5.11) \qquad \frac{\partial p}{\partial n}(ih,(J-1/2)h,n\Delta t) \sim \frac{1}{2}\left(\frac{1}{h}\left(p^{n+1}_{iJ} - p^{n+1}_{i,J-1}\right) + \frac{1}{h}\left(p^{n-1}_{iJ} - p^{n-1}_{i,J-1}\right)\right).$$

For the second equation on Γ in (5.4) with second order time derivatives, the approximation is centered in time at instant n. Thus the first and second time derivatives are approximated as in the previous scheme (5.6)-(5.8) except here we take the average of the centered difference at (ih,Jh) and the centered difference at (ih,(J-1)h) :

$$(5.12) \qquad \frac{\partial u}{\partial t}(ih,(J-1/2)h,n\Delta t) \sim \left(\frac{1}{2\Delta t}\left(u^{n+1}_i - u^{n-1}_i\right)\right)$$

$$(5.13) \qquad \frac{\partial^2 u}{\partial t^2}(ih,(J-1/2)h,n\Delta t) \sim \frac{1}{\Delta t^2}\left(u^{n+1}_i - 2u^n_i + u^{n-1}_i\right),$$

where u^n_i is either $(p^n_{iJ}+p^n_{iJ-1})/2$ or φ^n_{pi}, and for the second derivative in x we take the average of the centered difference at (ih,Jh) and at (ih,(J-1)h) and then we average this average calculated at instant n+1 and that calculated at instant n-1 :

$$(5.14) \qquad \frac{\partial^2 u}{\partial x^2}(ih,(J-1/2)h,n\Delta t) \sim \frac{1}{2}\left(\frac{1}{h^2}\left(u^{n+1}_{i+1} - 2u^{n+1}_i + u^{n+1}_{i-1}\right) + \frac{1}{h^2}\left(u^{n-1}_{i+1} - 2u^{n-1}_i + u^{n-1}_{i-1}\right)\right).$$

The use of instant n+1 in (5.11) and (5.14) makes this scheme implicit on Γ. However, in the calculations we have done, this scheme has remained stable even for weakly well posed problems corresponding to an impedance operator in Σ^W_a, cf. Section 6.

In fact we have the following theorem which we shall prove for the simplest case in which we can encounter a weakly well posed problem ; i.e. when $\hat{Z}$ is of the following form :

$$\hat{Z}(k,\omega) = \frac{1}{c}\left(1 - \frac{ak^2}{\dfrac{2}{\omega}}\right),$$

and the boundary condition from (5.4)) on Γ is of the form

$$\frac{\partial p}{\partial z} = -\frac{1}{c}\left(\frac{\partial p}{\partial t} - \frac{\partial \varphi}{\partial t}\right)$$

(5.15)

$$\frac{\partial^2 \varphi}{\partial t^2} = a\frac{\partial^2 p}{\partial x^2}$$

or equivalently

(5.16) $\quad \dfrac{\partial^2 p}{\partial t^2} + c\dfrac{\partial^2 p}{\partial t\partial z} - a\dfrac{\partial^2 p}{\partial x^2} = 0.$

(We have weak stability if c and a are positive, strong stability if in addition $a < 1$).

Theorem 5.1.

Suppose we have the following problem :

$$\frac{\partial^2 p}{\partial t^2} - \Delta p = f \qquad\qquad \text{for } z < 0$$

(5.17) $\quad \dfrac{\partial^2 p}{\partial t^2} + c\dfrac{\partial^2 p}{\partial t\partial z} - a\dfrac{\partial^2 p}{\partial x^2} = 0 \qquad \text{for } z = 0$

$$p = \frac{\partial p}{\partial t} = 0 \qquad\qquad \text{for } z < 0 \text{ and } t = 0$$

with a and c positive real constants. Then for $1/4 < \theta \le 1/2$, the numerical scheme

(5.18) $\quad h \quad p_{ij}^n = \dfrac{1}{\Delta t^2}\left(p_{ij}^{n+1} - 2p_{ij}^n + p_{ij}^{n-1}\right) - \dfrac{1}{h^2}\left(p_{ij+1}^n + p_{i,j-1}^n + p_{i+1,j}^n + p_{i-1,j}^n - 4p_{ij}^n\right) = f_{ij}^n \quad j = -\infty,\dots,-1$

(5.19) $\quad \dfrac{1}{\Delta t^2}\left(p_{i,-1/2}^{n+1} - 2p_{i,-1/2}^n + p_{i,-1/2}^{n-1}\right) + \dfrac{c}{2\Delta t}\left(\dfrac{p_{i0}^{n+1} - p_{i,-1}^{n+1}}{h} - \dfrac{p_{i0}^{n-1} - p_{i,-1}^{n-1}}{h}\right)$

$$- a\left(\theta D_x^2 p_{i,-1/2}^{n+1} + (1-2\theta)D_x^2 p_{i,-1/2}^n + \theta D_x^2 p_{i,-1/2}^{n-1}\right) = 0$$

(5.20) $\quad p_{ij}^0 = p_{ij}^1 = 0 \qquad j = -\infty,\dots,-1,$

where

(5.21) $\quad p_{i,j-1/2}^n = 1/2\left(p_{ij}^n + p_{i,j-1}^n\right)$

and

(5.22) $\quad D_x^2 p_{ij}^n = \left(p_{i+1,j}^n - 2p_{ij}^n + p_{i-1,j}^n\right)/h^2,$

is weakly in the sense that $\|p_h^n\| = \{\sum_{j=-\infty}^0 \sum_{i=-\infty}^\infty (p_{ij}^n)^2 h^2\}^{1/2}$ may be bounded in termes of a constant dependent on n and on f and its derivatives up through second order.

We point out that for $\theta=1/2$ the numerical scheme given by (5.18)-(5.19) is equivalent to that described by (5.5),(5.10)-(5.14) for the boundary condition (5.15) or (5.16).

Proof :

Define $\mathcal{L}_h$ to be the discrete analogue of the operator $\mathcal{L}$ given by

$$\mathcal{L}(p) = \frac{\partial^2 p}{\partial t^2} + c\,\frac{\partial p}{\partial t \partial z} - a\,\frac{\partial^2 p}{\partial x^2}$$

for the above difference scheme :

(5.23) $$\mathcal{L}_h(p^n_{ij}) = \frac{1}{\Delta t^2}\left(p^{n+1}_{i,j-1/2} - 2p^n_{i,j-1/2} + p^{n-1}_{i,j-1/2}\right) + \frac{c}{2\Delta t}\left(\frac{p^{n+1}_{ij} - p^{n+1}_{i,j-1}}{h} - \frac{p^{n-1}_{ij} - p^{n-1}_{i,j-1}}{h}\right)$$

$$- a\left(\theta D^2_x p^{n+1}_{i,j-1/2} + (1-2\theta)D^2_x p^n_{i,j-1/2} + \theta D^2_x p^{n-1}_{i,j-1/2}\right)$$

Now if p^n_{ij} is the solution of (5.18)-(5.20) then $v^n_{i,j-1/2}$ given by

$$v^n_{i,j-1/2} = \mathcal{L}_h(p^n_{ij}) \qquad j = -\infty,\ldots,0 \,;\, i = -\infty,\ldots,\infty \,;\, n = 2,\ldots\infty$$

(5.24) $$v^0_{i,j-1/2} = f^0_{i,j-1/2} - \frac{1}{2}\frac{c\Delta t}{h}(f^0_{ij} - f^0_{i,j-1}) - a\Delta t^2\,\theta D^2_x f^0_{i,j-1/2}$$

$$v^1_{i,j-1/2} = f^1_{i,j-1/2} + \frac{1}{2}\frac{c\Delta t}{h}(f^1_{ij} - f^1_{i,j-1}) - a\Delta t^2\,\theta D^2_x f^1_{i,j-1/2}$$

may also be defined as the solution of the Dirichlet problem

$$_h v^n_{i,j-1/2} = \mathcal{L}_h(f^n_{ij}) \qquad i = -\infty,\ldots,\infty \,;\, j = -\infty,\ldots,-1 \,;\, n = 2,\ldots,\infty$$

$$v^n_{i,-1/2} = 0 \qquad i = -\infty,\ldots,\infty \,;\, n = 2,\ldots,\infty$$

(5.25) $$v^0_{i,j-1/2} = f^0_{i,j-1/2} - \frac{1}{2}\frac{c\Delta t}{h}(f^0_{ij} - f^0_{i,j-1}) - a\Delta t^2\,\theta D^2_x f^0_{i,j-1/2}$$

$$v^1_{i,j-1/2} = f^1_{i,j-1/2} + \frac{1}{2}\frac{c\Delta t}{h}(f^1_{ij} - f^1_{i,j-1}) - a\Delta t^2\,\theta D^2_x f^1_{i,j-1/2}$$

and we have the estimate

(5.26) $$\|\bar{v}^n_h\| \leq \text{const}\left(\|\bar{v}^0_h\| + \|\bar{v}^1_h\| + \|\overline{\mathcal{L}_h f^n}\|\right) \leq V^n,$$

for some constant V^n depending on the derivatives of f up through second order, in the norm

(5.27) $$\|q^n_h\| = \left\{\sum_{j=-\infty}^{0}\sum_{i=-\infty}^{\infty}\left(q^n_{i,j-1/2}\right)^2 h^2\right\}^{1/2}.$$

Then p^n_{ij} is the solution of

(5.28) $$\mathcal{L}_h(p^n_{ij}) = v^n_{i,j-1/2} \qquad j = -\infty,\ldots,0 \,;\, i = -\infty,\ldots,\infty \,;\, n = 2,3\ldots$$

$$p^0_{ij} = 0 \qquad j = -\infty,\ldots,0 \,;\, i = -\infty,\ldots,\infty$$

$$p^1_{ij} = 0 \qquad j = -\infty,\ldots,0 \,;\, i = -\infty,\ldots,\infty$$

The idea of the proof, now following the development in [1] for the continuous paraxial approximation of the wave equation, is to estimate $\|p^n_h\|$ in terms of $\|v^n_h\|$. This could not be done if $\mathcal{L}_h$ were replaced by an operator $\widetilde{\mathcal{L}}_h$ corresponding to the explicit scheme (5.6)-(5.9) in the definition of v^n_{ij}.

To bound $\| \bar{p}^{\,n}_h \|$, we multiply (5.28) by
$$\left(p^{n+1}_{i,j-1/2} - p^{n-1}_{i,j-1/2} \right) / (2\Delta t),$$
sum on i and multiply by h to obtain

$$\frac{h}{2\Delta t} \sum_i \left[\left(\frac{p^{n+1}_{i,j-1/2} - p^{n}_{i,j-1/2}}{\Delta t} \right)^2 - \left(\frac{p^{n}_{i,j-1/2} - p^{n-1}_{i,j-1/2}}{\Delta t} \right)^2 \right]$$

$$+ c \sum_i \left[\left(\frac{p^{n+1}_{ij} - p^{n-1}_{ij}}{2\Delta t} \right)^2 - \left(\frac{p^{n+1}_{i,j-1} - p^{n-1}_{i,j-1}}{2\Delta t} \right)^2 \right]$$

$$- ah \sum_i D^2_x p^n_{i,j-1/2} \left(\frac{p^{n+1}_{i,j-1/2} - p^{n-1}_{i,j-1/2}}{2\Delta t} \right)$$

$$- a\theta h \sum_i D^2_x \left(p^{n+1}_{i,j-1/2} - 2p^{n}_{i,j-1/2} + p^{n-1}_{i,j-1/2} \right) \frac{p^{n+1}_{i,j-1/2} - p^{n-1}_{i,j-1/2}}{2\Delta t}$$

$$= h \sum_i v^n_{i,j-1/2} \left(\frac{p^{n+1}_{i,j-1/2} - p^{n-1}_{i,j-1/2}}{2\Delta t} \right).$$

We shall need some notation :

$$p^n_h = \left\{ p^n_{ij} \right\}_{\substack{i=-\infty,...,\infty \\ j=-\infty,...,0}} \qquad \bar{p}^{\,n}_h = \left\{ p^n_{i,j-1/2} \right\}_{\substack{i=-\infty,...,\infty \\ j=-\infty,...,0}}$$

$$p^n_{hj} = \left\{ p^n_{ij} \right\}_{i=-\infty,...,\infty}$$

$$\left(p^n_{hj}, q^m_{h\ell} \right) = h \sum_{i=-\infty}^{\infty} p^n_{ij} q^m_{i\ell} \qquad |p^n_{hj}| = \left(p^n_{hj}, p^n_{hj} \right)^{1/2}$$

(5.29)

$$\left(\left(p^n_h, q^m_h \right) \right) = h \sum_{j=-\infty}^{0} \left(p^n_{hj}, q^m_{hj} \right) \qquad \| p^n_h \| = \left(\left(p^n_h, p^n_h \right) \right)^{1/2}$$

$$d\left(p^n_{hj}, q^m_{h\ell} \right) = h \sum_{i=-\infty}^{\infty} \left(\frac{p^n_{ij} - p^n_{i-1,j}}{h}, \frac{q^m_{i\ell} - q^m_{i-1,\ell}}{h} \right)$$

$$d\left(\left(p^n_h, q^m_h \right) \right) = h \sum_{j=-\infty}^{0} d\left(p^n_{hj}, q^m_{hj} \right).$$

Now we sum over j, j = $-\infty$, 0 and multiply by h. Standard calculations lead to

$$\frac{1}{2\Delta t} \left\{ \left\| \frac{\bar{p}^{\,n+1}_h - \bar{p}^{\,n}_h}{\Delta t} \right\|^2 - \left\| \frac{\bar{p}^{\,n}_h - \bar{p}^{\,n-1}_h}{\Delta t} \right\|^2 \right\} + c \left| \frac{p^{n+1}_{h0} - p^{n-1}_{h0}}{2\Delta t} \right|^2$$

$$+ \frac{a}{2\Delta t} \left\{ d\left(\left(\frac{\bar{p}^{\,n+1}_h + \bar{p}^{\,n}_h}{2}, \frac{\bar{p}^{\,n+1}_h + \bar{p}^{\,n}_h}{2} \right) \right) - d\left(\left(\frac{\bar{p}^{\,n}_h + \bar{p}^{\,n-1}_h}{2}, \frac{\bar{p}^{\,n}_h + \bar{p}^{\,n-1}_h}{2} \right) \right) \right\}$$

$$+ \frac{a\Delta t}{2}(\theta - 1/4) \left\{ d\left(\left(\frac{\overline{P}_h^{n+1} - \overline{P}_h^n}{\Delta t}, \frac{\overline{P}_h^{n+1} - \overline{P}_h^n}{\Delta t}\right)\right) - d\left(\left(\frac{\overline{P}_h^n - \overline{P}_h^{n-1}}{\Delta t}, \frac{\overline{P}_h^n - \overline{P}_h^{n-1}}{\Delta t}\right)\right) \right\}$$

$$= \left(\left(\overline{v}_h^n, \frac{\overline{P}_h^{n+1} - \overline{P}_h^{n-1}}{2\Delta t}\right)\right).$$

Define the energy form :

$$(5.30) \qquad E^{n+1/2} = \left\| \frac{\overline{P}_h^{n+1} - \overline{P}_h^n}{\Delta t} \right\|^2 + a\, d\left(\left(\frac{\overline{P}_h^{n+1} + \overline{P}_h^n}{2}, \frac{\overline{P}_h^{n+1} + \overline{P}_h^n}{2}\right)\right) + a\,\Delta t^2(\theta - 1/4)\, d\left(\left(\frac{\overline{P}_h^{n+1} - \overline{P}_h^n}{\Delta t}, \frac{\overline{P}_h^{n+1} - \overline{P}_h^n}{\Delta t}\right)\right)$$

We have :

$$\frac{1}{2\Delta t}\left(E^{n+1/2} - E^{n-1/2}\right) + c\left| \frac{P_{h0}^{n+1} - P_{h0}^{n-1}}{2\Delta t} \right|^2 = \left(\left(\overline{v}_h^n, \frac{\overline{P}_h^{n+1} - \overline{P}_h^{n-1}}{2\Delta t}\right)\right).$$

Summing over n, n=1,...,N and multiplying by $2\Delta t$ we obtain :

$$E^{N+1/2} + 2c \sum_{n=1}^{N} \left| \frac{P_{h0}^{n+1} - P_{h0}^{n-1}}{2\Delta t} \right|^2 \Delta t = E^{1/2} + 2\sum_{n=1}^{N} \left(\left(\overline{v}_h^n, \frac{\overline{P}_h^{n+1} - \overline{P}_h^{n-1}}{2\Delta t}\right)\right) \Delta t.$$

If $\theta \geq 1/4$, we have :

$$\left\| \frac{\overline{P}_h^{N+1} - \overline{P}_h^N}{\Delta t} \right\|^2 \leq E^{N+1/2} \leq 2\sum_{n=1}^{N} \left(\frac{\|\overline{v}_h^{n+1}\| + \|\overline{v}_h^n\|}{2} \right) \left\| \frac{\overline{P}_h^{n+1} - \overline{P}_h^n}{\Delta t} \right\| \Delta t.$$

Then applying a discrete Grönwall lemma we obtain :

$$(5.31) \qquad \left\| \frac{\overline{P}_h^{N+1} - \overline{P}_h^N}{\Delta t} \right\| \leq 2\sum_{n=1}^{N+1} \|\overline{v}_h^n\| \Delta t,$$

and

$$(5.32) \qquad \|\overline{P}_h^{N+1}\| \leq 2N\Delta t \sum_{n=1}^{N+1} \|\overline{v}_h^n\| \Delta t.$$

Thus we have bounded $\|\overline{p}_h^N\|$ in terms of $t^N \sum_{n=1}^{N} \|v_h^n\| \Delta t$, where $\|\overline{p}_h^n\| = \sum_j \sum_i \|p_{i,j-1/2}^n\|^2 h^2)^{1/2}$. In order to obtain a bound for $\|p_h^n\| = \sum_j \sum_i \|p_{ij}^n\|^2 h^2)^{1/2}$ it suffices to bound $\|D_z p_h^n\|$ where by $D_z p_h^n$ we mean

$$(5.33) \qquad D_z p_h^n = \left\{ \frac{p_{ij}^n - p_{i,j-1}^n}{h} \right\}_{\substack{i=-\infty,...,\infty \\ j=-\infty,...,0}}$$

Towards this end we return to (5.28) and multiply by the following discrete analogue of $(\partial p/\partial t) + c(\partial p/\partial z)$:

$$\frac{p_{i,j-1/2}^{n+1} - p_{i,j-1/2}^{n-1}}{2\Delta t} + c\left(\theta \frac{p_{ij}^{n+1} - p_{i,j-1}^{n+1}}{h} + (1-2\theta) \frac{p_{ij}^n - p_{i,j-1}^n}{h} + \theta \frac{p_{ij}^{n-1} - p_{i,j-1}^{n-1}}{h} \right).$$

After summing on i and j and multiplying by h^2 we obtain :

$$\frac{1}{2\Delta t}\left(\widetilde{E}^{n+1/2} - \widetilde{E}^{n-1/2}\right) + (\theta - 1/4)\frac{c\Delta t^2}{2}\left|\frac{p_{h0}^{n+1} - 2p_{h0}^n + p_{h0}^{n-1}}{\Delta t^2}\right|^2$$

$$+ ac\, d\left(\theta p_{h0}^{n+1} + (1-2\theta)\, p_{h0}^n + \theta p_{h0}^{n-1},\ \theta p_{h0}^{n+1} + (1-2\theta)\, p_{h0}^n + \theta p_{h0}^{n-1}\right)$$

$$= \left(\left(v_h^n\ ,\ \frac{\overline{p}_h^{n+1} - \overline{p}_h^{n-1}}{2\Delta t} + c\left(\theta\,\frac{p_{ij}^{n+1} - p_{i,j-1}^{n+1}}{h} + (1-2\theta)\,\frac{p_{ij}^n\, p_{i,j-1}^n}{h} + \theta\,\frac{p_{ij}^{n-1} - p_{i,j-1}^{n-1}}{h}\right)\right)\right)$$

where the energy $\widetilde{E}^{n+1/2}$ is given by :

$$\widetilde{E}^{n+1/2} = \left\|\frac{\overline{p}_h^{n+1} - \overline{p}_h^n}{\Delta t} + c\left(\frac{D_z p_h^{n+1} + D_z p_h^n}{2}\right)\right\|^2$$

$$+ (\theta - 1/4)\, c^2\, \|D_z p_h^{n+1} - D_z p_h^n\|^2$$

$$+ a\, d\left(\left(\frac{\overline{p}_h^{n+1} + \overline{p}_h^n}{2},\ \frac{\overline{p}_h^{n+1} + \overline{p}_h^n}{2}\right)\right)$$

$$+ (\theta - 1/4)\, a\Delta t^2\, d\left(\left(\frac{\overline{p}_h^{n+1} - \overline{p}_h^n}{\Delta t},\ \frac{\overline{p}_h^{n+1} - \overline{p}_h^n}{\Delta t}\right)\right)$$

Thus for $\theta \geq 1/4$ we have :

$$\left\|\frac{\overline{p}_h^{N+1} - \overline{p}_h^N}{\Delta t} + c\,\frac{D_z p_h^{N+1} + D_z p_h^N}{2}\right\|^2$$

$$+ (\theta - 1/4)\, c^2\, \|D_z p_h^{N+1} - D_z p_h^N\|^2$$

$$\leq E^{N+1/2}$$

$$\leq E^{1/2} + 2\sum_{n=1}^{N}\left(\left(\overline{v}_h^n\ ,\ \frac{\overline{p}_h^{n+1} - \overline{p}_h^{n-1}}{2\Delta t} + \frac{1}{4}c\left(D_z p_h^{n+1} + 2D_z p_h^n + D_z p_h^{n-1}\right)\right)\right)\Delta t$$

$$+ 2\sum_{n=1}^{N}\left(\left(\overline{v}_h^n\ ,\ (\theta - 1/4)\, c\left(D_z p_h^{n+1} - 2D_z p_h^n + D_z p_h^{n-1}\right)\right)\right)\Delta t$$

$$\leq 2\sum_{n=1}^{N}\left(\frac{\|\overline{v}_h^n\| + \|\overline{v}_h^{n+1}\|}{2}\right)\left\|\frac{\overline{p}_h^{n+1} - \overline{p}_h^n}{\Delta t} + c\,\frac{D_z p_h^{n+1} + D_z p_h^n}{2}\right\|\Delta t$$

$$+ 2\sqrt{\theta - 1/4}\,2\sum_{n=1}^{N}\left(\frac{\|\overline{v}_h^n\| + \|\overline{v}_h^{n+1}\|}{2}\right)\sqrt{\theta - 1/4}\ c\,\|D_z p_h^{n+1} - D_z p_h^n\|\,\Delta t.$$

Again using a discrete Grönwall lemma we obtain

$$(5.35) \quad \left\{\left\|\frac{\overline{p}_h^{N+1} - \overline{p}_h^N}{\Delta t} + c\,\frac{D_z p_h^{N+1} + D_z p_h^N}{2}\right\|^2 + (\theta - 1/4)\, c^2\, \|D_z p_h^{N+1} - D_z p_h^N\|^2\right\}^{1/2}$$

$$\leq 2 \left(1+\sqrt{\theta-1/4}\ \right) \sum_{n=1}^{N} \| \overline{v}_h^{-n} \| \, \Delta t.$$

Now (5.31) together with (5.35) imply that for $1/4 < \theta \leq 1/2$:

$$\| D_z p_h^N \| \leq \left(3+\sqrt{\theta-1/4}\ \right) \sum_{n=1}^{N} \| \overline{v}_h^{-n} \|,$$

and the theorem follows.

6 - Numerical results

In the numerical experiments presented here we compare the results obtained in Ω_F using an approximate impedance operator Z_a with those obtained by solving the transmission problem in $\Omega_F \cup \Omega_S$. We look at two cases. In both Ω_S is a homogeneous acoustic medium. In the first, the velocity C_S in Ω_S is 1/2 so that $Z_N \in \Sigma^S{}_a$ for all N and the initial boundary value problem in Ω_F is strongly well posed. (The velocity in Ω_F is $C_F=1$). In the second, $C_S=5$ so that for $N \geq 2$, Z_N is in $\Sigma^W{}_a$ but not in $\Sigma^S{}_a$ and the initial boundary value problem is only weakly well posed.

To obtain the reference solution, the so called exact solution, in $\Omega_F \cup \Omega_S$ we place the lower boundary as well as the lateral boundaries of the domain of calculation sufficiently far away from the source that they don't affect the numerical solution. (The upper boundary is the free boundary Γ_F). Then the simulation in Ω_F using an impedance operator Z_m is carried out by bringing the lower boundary up to Γ.

The source is a point source, the product of a Gaussian $g(t)$ in time and a Dirac $h(x,z)$ in space :

$$f(x,z,t) = h(x,z)g(t)$$

$$g(t) = \begin{cases} \exp(-10(1-t_S)^2) & 0 < t < 2t_S \\ 0 & 2t_S < t \end{cases}$$

$$h(x,z) = \begin{cases} 10^4(1-r/r_S) & r < r_S \\ 0 & r_S \leq r \end{cases}$$

$$r = ((x-x_S)^2 + (z-z_S)^2)^{1/2}$$

This source is centered at a point $(x_S\text{-}z_S) \in \Omega_F$ at a distance 0.3 from Γ and has a radius $r_S=0.08$. It is active between instants $t=0$ and $t=2t_S=0.2$.

The grid is a regular grid with $\Delta x=\Delta z=0.01$ which corresponds to roughly 10 points per wave length effectively represented. The time step Δt is $\Delta t = \overline{c}\ h/2$ where $\overline{c}$ is the larger of the two velocities c_F and c_S.

Figures 4-7 correspond to the case $c = 1/2$. The surface shown represents the pressure p(x,z,t) in Ω_F at instant t=0.75. In addition to the exact solution we show the solutions obtained in Ω_F with the impedance operators Z_1, Z_2 and Z_3 given by (4.16), (4.17) and (4.18). The explicit numerical scheme (5.6)-(5.9) was used for the calculations on Γ. There is no problem of stability. All three impedance operators yield solutions which are excellant approximations to the exact solution.

Figures 8-14 correspond to the case c=5. Again the pressure p(x,z,t) in Ω_F at t=0.75 is shown. The exact solution is given in *Figure 8* and that obtained with the impedance operator $Z_1 \in \Sigma^S{}_a$, (4.16), is shown in *Figure 9*. We see that stability considerations aside, the approximation of the true impedance is more difficult here than in the preceeding case. Besides the exact solution, there are two solutions obtained with the impedance operator Z_2 given by (4.17), one shown in *Figure 10* with the explicit numerical scheme on Γ, (5.6)-(5.9), and the other *Figure 11* with the implicit scheme on Γ, (5.10)-(5.13). In *Figure 10* we see typical indications of instability, and the energy curve exploded for this simulation.

In *Figure 11* the solution, while more reasonable than that of *Figure 10*, is not entirely precise. There is a larger reflection for larger angles of incidence than is correct and this reflection does propagate into the domain Ω_F. The energy shown in *Figure 13* increases linearly in time. These are both characteristics of weak stability. Whether this provides a better approximation of the measured solution than the solution obtained with an impedance operator in $\Sigma^S{}_a$ is likely to depend on the zone of observation or measure, cf. Section 7.

Figure 12 is obtained with an operator Z_a corresponding to $r_a(\theta)$ given by (4.21) for $\varepsilon=0$. In this case $r_a(1) = 1$ and while Z_a is not in $\Sigma^S{}_a$ there are no indications at instability ; the energy, *Figure 14*, decreases (cf.Remark 4.1).

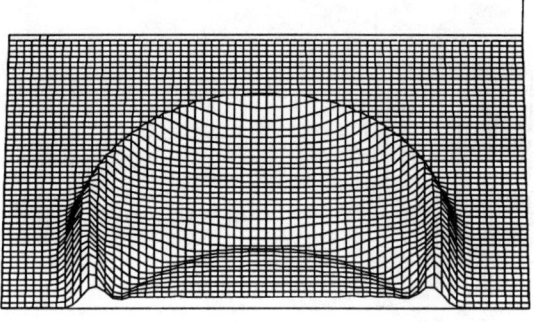

Fig. 4 Solution at t = 0.75
♦ cas c = 1/2
♦ exact solution
min = 0.02 , max = 1.43

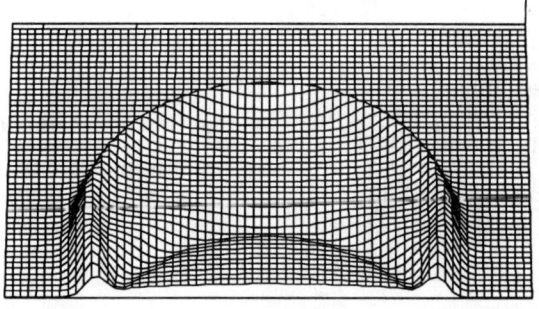

Fig. 5 Solution at t = 0.75
♦ cas c = 1/2
♦ solution with boundary
   condition of order 1
min = 0.04 , max = 1.43

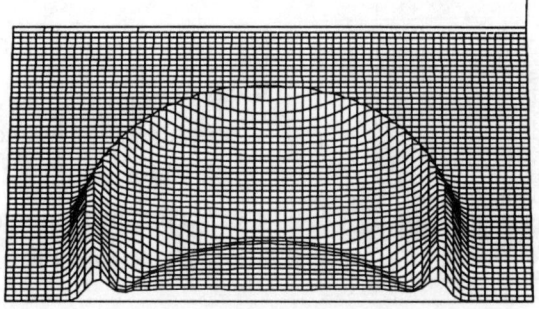

Fig. 6 Solution at t = 0.75
♦ cas c = 1/2
♦ solution with boundary
   condition of order 2
min = 0.02 , max = 1.43

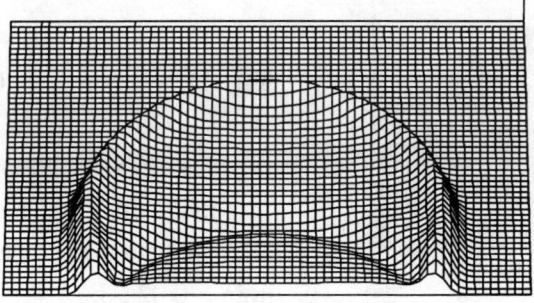

Fig. 7 Solution at t = 0.75
♦ cas c = 1/2
♦ solution with boundary
   condition of order 3
min = 0.02 , max = 1.43

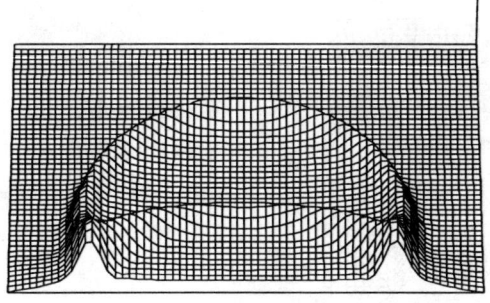

Fig. 8 Solution at t = 0.75
- cas c = 5
- exact solution
min = 0. , max = 2.48

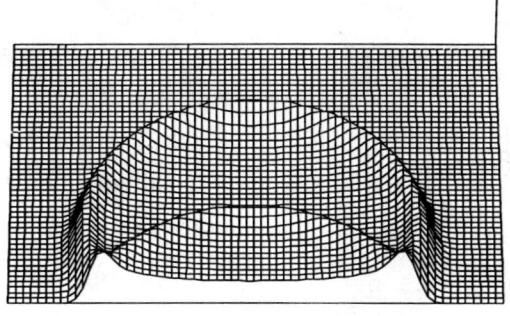

Fig. 9 Solution at t = 0.75
- cas c = 5
- solution with boundary
  condition of order 1
min = 0. , max = 2.02

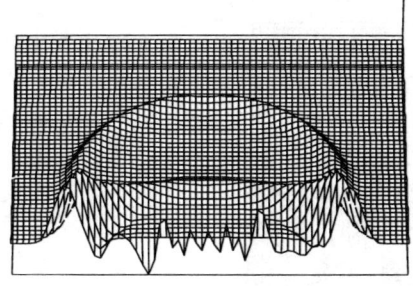

Fig. 10 Solution at t = 0.75
- cas c = 5
- solution with boundary
  condition of order 2
- explicit numerical scheme
min = -3.03 , max = 6.28

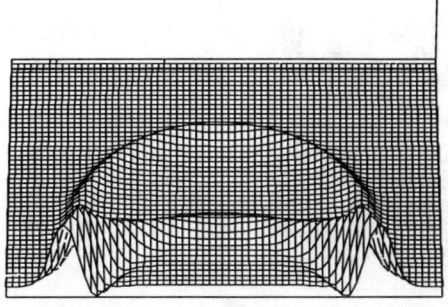

Fig. 11 Solution at t = 0.75
- cas c = 5
- solution with boundary
  condition of order 2
- implicit numerical scheme
min = 0.02 , max = 1.43

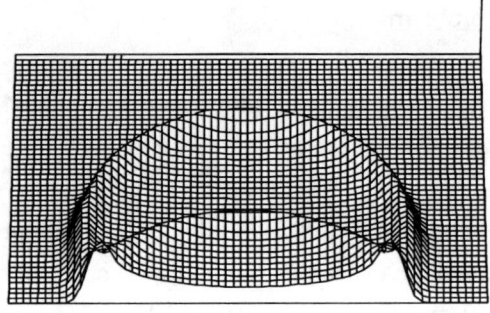

Fig. 12 Solution at t = 0.75

♦ cas c = 5

♦ solution with boundary condition associated with (4.21), ε = 0

♦ explicit numerical scheme

min = 0. , max = 2.49

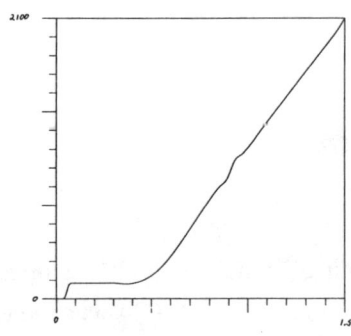

Fig. 13 Energy as a function of time

♦ cas c = 5

♦ boundary condition of order 1

♦ implicit numerical scheme

min = 0. , max = 2.02

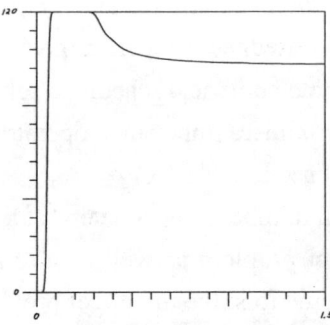

Fig. 14 Energy as a function of time

♦ cas c = 5

♦ boundary condition associated with (4.21), ε = 0

♦ explicit numerical scheme

7 - Conclusion : perspectives for the inverse problem

Problem (5.4) defines a mapping F associating to $Z \in \Sigma^S_a$ the solution p of (5.4). For a subset $\mathcal{O}$ of Ω_F, which we shall call the *domain of observation,* we denote by $F_\mathcal{O}$, the mapping F followed by restriction to $\mathcal{O}$: $F_\mathcal{O}(Z) = p|_\mathcal{O}$. The question which arises is can we determine Z from an observation of the solution p in $\mathcal{O}$; can we invert $F_\mathcal{O}$. More precisely, suppose that by an experiment we have observed the exact solution $p_\mathcal{O}(x,z,t)$ corresponding to the true impedance operator $Z \in \Sigma$ in the set $\mathcal{O}$ during the time interval [0,T]. Can we find the approximate impedance operator $Z_a \in \Sigma^N_a$ such that $F_\mathcal{O}(Z)$ best approximates, in the sense of least squares for example, the observation $p_\mathcal{O}$, where by Σ^N_a we denote the subclass of Σ^w_a consisting of those impedance operators whose complex impedance is of the form (3.4) for some $N' \leq N$. To find such a Σ_a it suffices to solve the following minimization problem. Find $Z^* \in \Sigma^W_a$ such that :

$$J(Z) \leq J(Z^*) \text{ for all } Z \in \Sigma^N_a$$

(7.1)

$$J(Z) = \left(\int_0^T \| F_\mathcal{O}(Z) - p_\mathcal{O} \|^2_{L^2(\mathcal{O})} \, dt \right)^{1/2}$$

Problem (7.1) is a problem of optimal control which we call the *inverse problem* . The function J is called the *criterion* or *cost function* . Problem (5.4) is called the *direct problem* or *equation of state* .

To solve (7.1) we need a minimization algorithm. All algorithms for the minimization of differentiable functions are iterative methods which make use of the gradient of J. To calculate J we solve a system of equations called the *adjoint equation* . This system is of the same type as is the state equation (5.4) and all the well posedness properties given in section 4 apply as well to the adjoint equation and we use the same numerical scheme to solve it.

Thus we have defined a space of approximate impedance operators which satisfies for the most part the desired properties (1)-(4) of section 3.

Each element of Σ^N_a is determined by a small number of parameters, at least for N small. Each element of Σ^N_a leads to a well posed direct problem as well as a to a well posed adjoint equation. We have defined a numerical scheme to solve the direct problem that is stable and simple to solve and which should work equally well for the adjoint equation.

For the inverse problem we can of course envisage several variations on the model problem (7.1) such as repeating the experiment with different source terms, or introducing a weight in the norm used to define the optimization criterion J. However we insist on pointing out the difficulty of the preposed inverse problem. The function F_Θ being very non linear, the fonctional we have to minimize is anything but quadratic. Consequently the problem of existence and especially uniqueness of the solution of the inverse problem is entirely open. Furthermore, the functional J being non convex we can expect the existence of local minima which renders even more arduous the task of the optimization algorithm.

Reference

[1] Bamberger A., Engquist B., Halpern L., Joly P., *Parabolic wave equation approximations in heterogeneous media,* SIAM J. Appl. Math. 48, 1988, pp.129-154.

[2] Engquist B., Majda, A., *"Absorbing boundary conditions for the numerical simulation of waves"* Math. of Comp., Vol.31, n° 139, pp.629-651, (1977).

[3] Engquist B., Majda, A., *"Radiation boundary conditions for acoustic and elastic wave calculations",* Comm. Pure Appl. Math., Vol.32, pp.313-357, (1979).

[4] Hersch R., *Mixed problems in several variables,* Jour. Math. Mech., 12, 1963, pp. 317-334.

[5] Higdon R.L., *Initial boundary value problems for hyperbolic systems,* SIAM Rev., Vol.28, 1986, pp. 177-217.

[6] Kreiss H.O., *Initial boundary value problems for hyperbolic systems,* Comm. Pure Appl. Math., Vol 23, 1970, pp.277-298.

[7] Kreiss H.-O., *Boundary conditions for hyperbolic differential equations,* Proc. Dundee Conference on Numerical Solution of Differential Equations, Lecture Notes in Mathematics 363, Springer-Verlag, New York, 1974, pp.64-74.

[8] Trefethen L.N. et Halpern L., *Well-posedness of one-way wave equations and absorbing boundary conditions.* Math. Comp., Vol.47, 1986, pp. 421-435.

CHAPTER 6

Wave Propagation by Step Marching

D. M. Pai*

Abstract

This paper presents a step-marching method for wave equation solution in in-homogeneous media. The solution obtained is the transmission signal. The method expands the solution in terms of plane-waves which during propagation are mixed by the medium inhomogeneity. Like the parabolic equation method, the present method is step-marching in space and thus very efficient. From comparisons to analytic solutions, the present method is more accurate than the parabolic equation method away from the major propagation direction.

1. Introduction

Wave equations in inhomogeneous media are non-separable and require numerical methods for solution. To reduce computation, one may employ approximate methods for solution. The parabolic equation method is such an approximation method.

The parabolic equation is a first order (in derivative) equation, derived from the

*Department of Electrical Engineering, University of Houston, Houston, TX 77204-4793

second order wave equation by assuming one-way wave propagation along a major propagation direction. The equation, being first order, with a given wave amplitude at an initial position, allows solution to be obtained by step-marching. In this way, the original boundary value problem is converted to a much easier (computationally) initial value problem. The parabolic equation method has been widely used in underwater acoustics [1],[2], electromagnetics [3] and exploration seismics [4].

This paper presents a new step-marching method, derived under totally different assumptions from the parabolic equation method. From comparisons to analytic solutions, the present method turns out to be just as accurate as the parabolic equation method along the major propagation direction, but proves to be more accurate away from the major propagation direction.

In the present method, the solution is expanded in terms of plane-waves. A full-wave continuation formula is constructed, and then, the reflection component of the solution is assumed to be small and the one-way wave approximation is applied to yield a step-marching solution. In step-marching, the plane-waves are mixed due to medium inhomogeneity.

2. Integral equation formulation

We use a 2-D scalar wave equation, to illustrate the method. Consider the wave equation

$$\left[\frac{\partial^2}{\partial x^2} + \frac{\partial^2}{\partial z^2} + \frac{\omega^2}{c^2(x,z)} \right] \psi(x,z) = 0 \tag{1}$$

with initial conditions $\psi(z_0)$ and $\psi'(z_0)$ (prime refers to the z-derivative) at z_0. The wave speed, $c(x,z)$, depends on both x and z, the horizontal and vertical coordinate, respectively. The discrete Fourier transform of equation (1) with respect to x is the matrix equation

$$\left[\frac{d^2}{dz^2} + \mathbf{K} + \omega^2 \mathbf{S}(z) \right] \Psi(z) = 0 \tag{2}$$

where

$$\mathbf{K} = \begin{pmatrix} -k_1^2 & & 0 \\ & \ddots & \\ 0 & & -k_N^2 \end{pmatrix}, \tag{3a}$$

and

$$\mathbf{S}(z) = \begin{pmatrix} s_0(z) & \cdots & s_{-(N-1)}(z) \\ \vdots & & \vdots \\ s_{N-1}(z) & \cdots & s_0(z) \end{pmatrix}. \tag{3b}$$

In the above, k is the horizontal wavenumber, with discretization interval $\Delta k = 2\pi/L$, defined by the horizontal length of the problem, L. The vector $\Psi(z)$ is the k-space solution vector truncated to N elements. The elements $s_n(z)$ are the Fourier components of $c^{-2}(x,z)$, $s_n = (1/L) \int c^{-2}(x) \exp(-in\Delta kx) dx$. Here, $S(z)$ is a Hermitian Toeplitz matrix. Equation (2) is a set of coupled ordinary differential equations. Let $S_0(z)$ denote the diagonal part of $S(z)$ and rewrite equation (2) as

$$\left[d^2/dz^2 + Q^2(z)\right] \Psi(z) = V(z)\Psi(z). \tag{4}$$

The left hand side is a decoupled system, defined by $Q^2(z) = K + \omega^2 S_0(z)$. All the coupling is lumped to the right hand side, defined by $V(z) = -\omega^2 [S(z) - S_0(z)]$, which is a measure of the lateral inhomogeneity at z. Treating the right hand side in terms of the Green's function of the left hand side, one can convert equation (2) into the integral equation

$$\Psi(z) = \Phi(z) + \int_{z_0}^{z} G(z,z')V(z')\Psi(z')dz', \tag{5}$$

where

$$\left[d^2/dz^2 + Q^2(z)\right] G(z,z') = \delta(z-z')I \tag{6}$$

with initial conditions $G(z,z') \equiv 0$ for $z \leq z'$, and

$$\left[d^2/dz^2 + Q^2(z)\right] \Phi(z) = 0 \tag{7}$$

with initial conditions $\Phi(z_0) = \Psi_0(z_0)$ and $\Phi'(z_0) = \Psi'_0(z_0)$. $G(z,z')$ is a diagonal matrix. It is readily seen that $\Psi(z)$ as given by equation (5) satisfies equation (1) and initial conditions at z_0.

3. Solution continuation: the propagation matrix

Equation (5) can be used for solution continuation from z_0 through a series of small steps to any depth. Consider the solution at z_1, a small step Δz from z_0. Since Δz is small, we make the following assumptions $G(z_1,z') \simeq G(z_1,z_0)$, $\Psi(z') \simeq \Psi(z_0)$ and $V(z') \simeq V(z_0)$. As a result,

$$\Psi(z_1) = \Phi(z_1) + G(z_1,z_0)V(z_0)\Psi(z_0)\Delta z. \tag{8}$$

Furthermore, since Δz is small, the background $Q^2(z)$ can be taken to be $Q^2(z_0)$, a constant. Let q_n^2 denote the components of $Q^2(z_0)$. It follows that, from z_0 to z_1, the components of Φ is a linear combination of up and down waves,

$$\phi_n(z) = u_n(z_0)e^{-iq_n(z-z_0)} + d_n(z_0)e^{iq_n(z-z_0)}, \tag{9}$$

where the constants, $u_n(z_0)$ and $d_n(z_0)$, are the up- and down-wave amplitudes at z_0 (z pointing downward). Similarly, the components of the Green's matrix (which is diagonal) are given by

$$g_n(z_1, z_0) = (-1/2iq_n)e^{-iq_n(z_1-z_0)} + (1/2iq_n)e^{iq_n(z_1-z_0)}, \tag{10}$$

also a combination of up and down waves. After a substitution of $\phi_n(z)$ and $g_n(z_1, z_0)$ into equation (8) and a separation of $\Psi(z)$ to up and down waves, $\Psi(z) = \mathbf{u}(z) + \mathbf{d}(z)$ [where $\mathbf{u}(z)$ and $\mathbf{d}(z)$ are respectively the up and down wave vectors], one obtains

$$\begin{aligned}\begin{pmatrix} \mathbf{u}(z_1) \\ \mathbf{d}(z_1) \end{pmatrix} &= \begin{pmatrix} \mathbf{P}_u(z_1, z_0) & 0 \\ 0 & \mathbf{P}_d(z_1, z_0) \end{pmatrix} \begin{pmatrix} \mathbf{u}(z_0) \\ \mathbf{d}(z_0) \end{pmatrix} \\ &+ \begin{pmatrix} \mathbf{G}_u(z_1, z_0)\mathbf{V}(z_0) & \mathbf{G}_u(z_1, z_0)\mathbf{V}(z_0) \\ \mathbf{G}_d(z_1, z_0)\mathbf{V}(z_0) & \mathbf{G}_d(z_1, z_0)\mathbf{V}(z_0) \end{pmatrix} \begin{pmatrix} \mathbf{u}(z_0) \\ \mathbf{d}(z_0) \end{pmatrix} \Delta z, \end{aligned} \tag{11}$$

where $\mathbf{P}_u(z_1, z_0)$, $\mathbf{P}_d(z_1, z_0)$, $\mathbf{G}_u(z_1, z_0)$ and $\mathbf{G}_d(z_1, z_0)$ are all diagonal matrices, with elements $e^{-iq_n(z_1-z_0)}$, $e^{iq_n(z_1-z_0)}$, $e^{-iq_n(z_1-z_0)}/-2iq_n$ and $e^{iq_n(z_1-z_0)}/2iq_n$, respectively.

An interpretation of equation (11) is given in Figure 1. As shown in the left panel, upon downward continuation, a single component of $\mathbf{u}$ - i.e. an up k-wave - is converted to three groups of waves: $\mathbf{P}_u\mathbf{u}$ converts the up k-wave to the same up k-wave multiplied by a phase factor, whereas $\mathbf{G}_u\mathbf{V}\mathbf{u}\Delta z$ converts the up k-wave to a spectrum of up k'-waves ($k' \neq k$) and $\mathbf{G}_d\mathbf{V}\mathbf{u}\Delta z$ converts the up k-wave to a spectrum of down k'-waves. The spreading of k to k', or the non-conservation of k, is a consequence of the lateral inhomogeneity. When the medium is laterally homogeneous, as illustrated in the right panel of Figure 1, only the term $\mathbf{P}_u\mathbf{u}$ remains and eack k-wave can be continued by itself. This interpretation leads to referring the method as a "plane-wave layer interaction" (PLI) method.

4. One-way wave approximation

Suppose we are only interested in the direct transmission signal and assume that back scatter is weak (one-way wave approximation), then, by ignoring the up-waves in equation (11), the propagation of a downward incident wave may be given by

$$\mathbf{d}(z_1) = \mathbf{P}_d(z_1, z_0)\mathbf{d}(z_0) + \mathbf{G}_d(z_1, z_0)\mathbf{V}(z_0)\mathbf{d}(z_0)\Delta z. \tag{12}$$

This solution requires only matrix-vector multiplications and thus can be very efficient in computation. The solution is step-marching in z.

5. The parabolic equation method

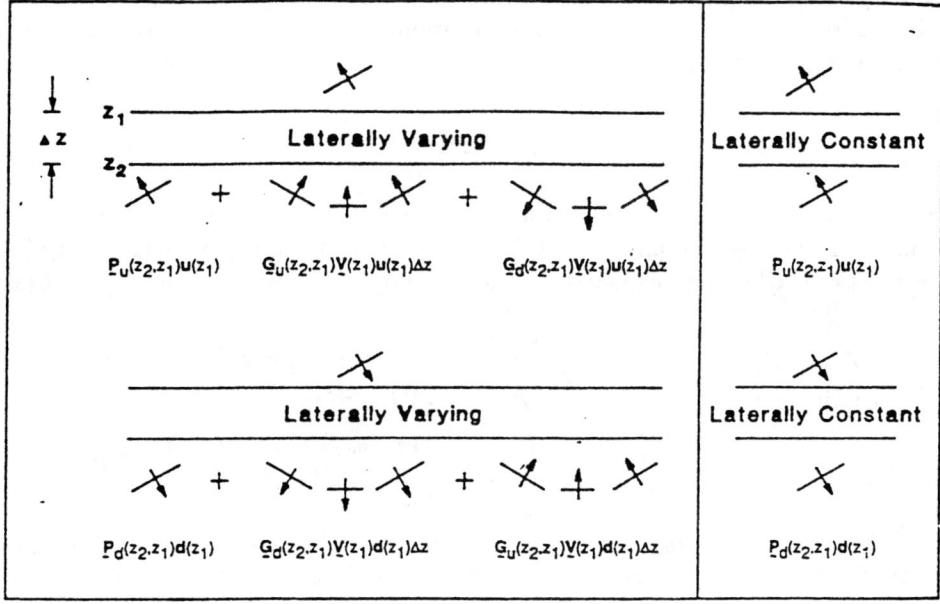

Figure 1   A graphical illustration of the planewave layer interaction described by equation (11).

For comparison, the parabolic equation is briefly presented below. The derivation is standard . Assuming the wave propagates mainly along the z-axis, then, by writing

$$\psi(x, z) = \exp(iK_0 z)\phi(x, z) \qquad (13)$$

where the propagation constant $K_0 = \omega/c_0$ is defined with respect to a constant background velocity, c_0, and by substituting the above into equation (1), one may derive

$$2iK_0 \frac{\partial\phi(x, z)}{\partial z} + \frac{\partial^2\phi(x, z)}{\partial x^2} + (K^2(x, z) - K_0^2)\phi(x, z) = 0. \qquad (14)$$

In the derivation, the $\partial^2\phi(x, z)/\partial z^2$ term has been assumed to be small compared to $2iK_0\partial\phi(x, z)/\partial z$ and has been ignored. Equation (14) is first order in z and is known as the parabolic equation. The first order derivative in z allows for solution by step-marching in z, given the incident wave amplitude at some initial depth.

6. Examples

To examine the method, plane-wave scattering by a circular cylinder is calculated and compared to analytical solutions. Let the incident plane-wave be from the top along the z-axis (k_{in}=0). (The parameters below are given in specific units, but may be scaled freely to other regimes.) Let ω be 500/sec (or frequency 79.6 Hz) and wave speed in host be 5 km/sec (or host wavelength 62.8 m). Let the cylinder radius be 100 m, simulating a medium speed variation at a length scale similar to wavelength. Three cylinder wave speeds are considered: 6 km/sec, 7 km/sec and 8 km/sec. Analytical solutions are obtained across a horizontal line from x=-1600 m to x=1600 m (L=3200 m) located at z=150 m (origin defined by center of the cylinder). The analytical solutions are then Fourier transformed to plane-wave components and are shown in Figure 2 as solid lines (with unit $\Delta k = 2\pi/L$), with figures from top to bottom corresponding to increasing cylinder wave speed. Since $k_{in} = 0$, the k-spectrum is symmetric and only positive k components are shown. Numerical solutions are obtained using $N = 149$ and are shown as squares in Figure 2. The left panel shows the results of the PE method, whereas the right panel shows those of the PLI method. For both methods, the agreement is good for low velocity contrast. As the velocity contrast increases, the agreement deteriorates, showing the breaking down of the one-way wave approximation. The two methods produce similar results at lower k values corresponding to lower scattering angles. However, the PLI method performs better than the PE method at higher k values corresponding to larger scattering angles. The PE method is consistently poor at higher scattering angles (for all three cases). This poor performance is not surprising, since the parabolic equation assumption does not apply there.

An interesting example is wave propagation through random media as calculated by using this method. Shown in Figure 3 is the transmission of a plane-wave pulse propagating through a medium consisting of a top uniform layer, a middle layer of randomly varying velocities, and a bottom uniform layer. The first diagram is the velocity profile of the three layers. The top and bottom layer has a constant velocity of 4.5 km/sec. The middle layer has a velocity distribution given by a normal-curve random number generator, with a mean of 3.0 km/sec and a standard deviation of 0.4 km/sec. The other three diagrams are the wave profile at three time instants: before, during, and after the pulse propagating through the middle layer. This example is calculated in the following way. Given the incident pulse, the wave field at $z = 0$ across all x is known as a function of time. The incident pulse is chosen to be $s(t) = t\exp(-\alpha t^2)$ where the pulse width $\Delta = (2/\alpha)^{1/2}$ (the width from the low peak to the high peak) is chosen to be 0.01 sec. The corresponding frequency spectrum is $s(\omega) = (-i\omega/\alpha)(\pi/4\alpha)^{1/2}\exp(-\omega^2/4\alpha)$, which has a dominant frequency at $\omega_0 = (2\alpha)^{1/2} = 2/\Delta$, or 200/sec. For each frequency component, the wave field at $z = 0$ is downward continued to all z by using the PLI method. The solution (at all z) in the frequency domain is then Fourier transformed to the time domain. The calculation is done by using $N = 50$ and by using 100 frequency components. The extent of the

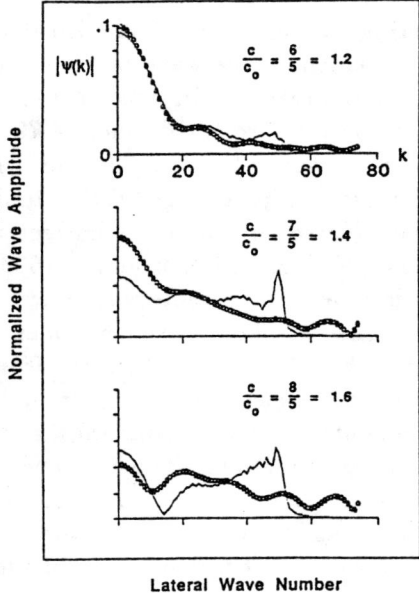 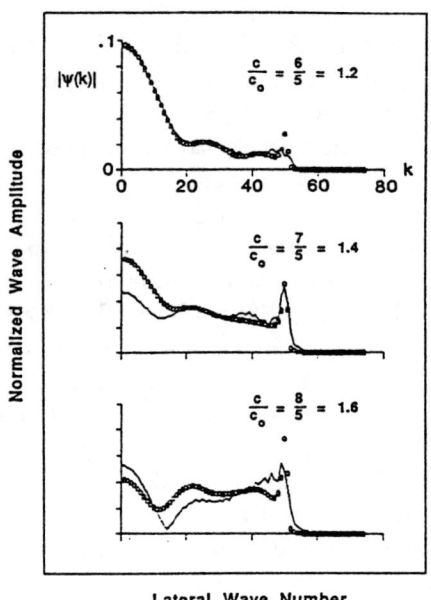

Figure 2   Planewave scattering by a circular cylinder: a comparison of analytical solution (the solid line) to numerical solutions (the squares) obtained by the PE method (left panel) and the PLI method (right panel).

model is 1.28 km on each side.

7. Conclusion

A one-way solution method has been introduced. The solution is expanded in terms of plane-waves which during propagation are continuously mixed by medium inhomogeneity. Similar to the parabolic equation method, the present method generates solutions by step-marching in space and thus can be very efficient. From comparisons to analytic solutions, the present method turns out to be just as accurate as the parabolic equation method along the major propagation direction, but proves to be more accurate away from the major propagation direction.

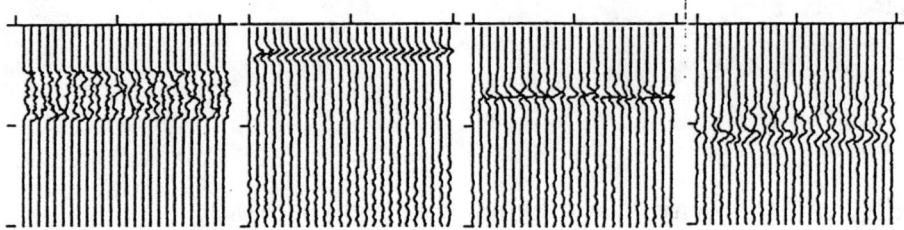

Figure 3   Plane-wave pulse propagation through random media.

References

[1] F.D. Tappert, *The parabolic approximation method*, in *Wave Propagation and Underwater Acoustics*, ed. J.B. Keller and J.S. Papadakis, Spring-Verlag, New York, 1977.

[2] J.A. DeSanto, *Theoretical methods in ocean acoustics*, in *Ocean Acoustics*, ed. J.A. DeSanto, Spring-Verlag, New York, 1979.

[3] H.V. Hitney, J.R. Richter, R.A. Pappert, K.D. Anderson and G.B. Baumgartner, *Tropospheric radio propagation assessment*, Proceedings IEEE, 73 (1985), pp. 265-283.

[4] J.F. Claerbout, *Imaging the Earth's Interior*, Blackwell Publication, 1985.

CHAPTER 7

Stability of One-Way Wave Equations as Absorbing
Boundary Conditions for the Wave Equation*

R. A. Renaut†

Abstract. Lindman first introduced the idea of finding an approximate one-way wave equation and using it as an absorbing boundary condition for the solution of the wave equation. Subsequently, Clayton and Engquist found the paraxial approximations by applying an alternative technique of approximation to the one-way wave equation. Halpern and Trefethen generalised the Clayton and Engquist method and derived whole families of approximations to the one-way wave equations. Here we show how to find finite difference approximations for any of the one-way wave equations by extending the ideas of Lindman. We analyse the stability of all these discretisations as absorbing boundary conditions for the solution of the wave equation with a standard five point difference stencil. A method for proving stability is given. In all cases we show that the bounds on the Courant number imposed by stability of the boundary condition are no more severe than the stability restriction for the interior scheme. The relevance of these results to problems where the difference scheme for the interior equation is not the five point stencil is also discussed.

1. Introduction. Absorbing boundary conditions have become a valuable tool in the modelling of the wave equation. In many geophysical problems the solution of the wave equation on an infinite domain is required. Computationally, a solution can only be found by imposing artificial boundaries at which all incident energy should be absorbed. In practice the ideal is not achieved but by careful choice of boundary condition, suited to the problem in question, the reflected energy may be minimised. Lindman [7] was the first to attempt to minimise reflection by solving an equation at the boundary which was derived from an approximation to the dispersion

---

*The work of the author was supported by National Science Foundation grant ASC 8812147 and an American Chemical Society Petroleum Research Fund grant 20681-G2.

†Department of Mathematics, Arizona State University, Tempe, AZ 85287–1804

relation of the one-way wave equation. Clayton and Engquist's method [1], however, has been more widely accepted.   They inverted the expression of Lindman and derived the low order paraxial approximations to the dispersion relation. Although their method is much simpler than Lindman's it is also less effective. Halpern and Trefethen [3] extended the work of Clayton and Engquist and derived whole families of approximations to the dispersion relation. Whereas the paraxial approximations have been fully tested, Halpern and Trefethen's [3] work only gives us the partial differential equations to be solved at the boundary without their discretisations in space and time.

In this paper we review the derivation of absorbing boundary conditions from approximations to the dispersion relation for the one-way wave equation.   We adapt the method of Lindman [1] to give a general discretisation of any of the absorbing boundary conditions introduced by Halpern and Trefethen [3].   Using Schur polynomials, we describe a general method for proving limits on the Courant number, μ, ratio of timestep to spacestep, for which the boundary condition is stable in conjunction with the standard five point discretisation of the wave equation. This method is used to present the bounds on μ for absorbing boundary conditions of orders up to order six with the discretisation proposed above. We show that the restriction on μ imposed by the boundary condition is in no case worse than that imposed by the five point difference operator. There do, however, exist generalised eigensolutions in some cases but, where they exist, the reflection coefficient is less than one. This means that these methods are potentially unstable but the numerical tests reported in Renaut [9] have not picked up the instability. The relevance of these results for other interior difference methods is also discussed and we argue that the results basically remain valid.

2. One-way wave equations. A one-way wave equation is an equation for which the solutions are waves that travel in one direction only. This compares to the second-order wave equation

$$u_{tt} = c^2(u_{xx} + u_{yy}), \tag{2.1}$$

$x > 0$, $y \in \mathbf{R}$, $t > 0$ and $u = u(x, y, t)$ for which the solutions are plane waves which travel in every direction in two dimensions. This can be seen by considering the dispersion relation for (2.1),

$$\omega^2 = c^2(\xi^2 + \eta^2) \tag{2.2}$$

which is obtained by substituting

$$u(x, y, t) = e^{i(\omega t + \xi x + \eta y)} \tag{2.3}$$

into (2.1), where ω is the frequency and ξ and η are wave numbers. A wave with wave numbers ξ and η travels at the velocity $c(-\xi/\omega, -\eta/\omega) = c(-\cos\theta, -\sin\theta)$, where θ is the angle measured counter-clockwise from the positive x-axis of the normal to the wave. Thus a wave which travels only to the left, for example, satisfies the dispersion relation

$$\xi = \frac{\omega}{c}\sqrt{1 - s^2} \tag{2.4}$$

where $s = \frac{\eta c}{\omega} = \sin\theta \in [-1, 1]$, and $\theta \in [-90°, 90°]$. Now suppose that we wish to solve (2.1) with an artificial boundary at $x = 0$. At the boundary we would like to allow only those waves which travel to the left and hence obey (2.4). But now equation (2.4) is not the dispersion relation of a partial differential equation because of the square root. Therefore practical one-way wave equations are found by using rational approximations to the square root function, or its inverse.

Let $r(s) = p_m(s)/q_n(s)$ be a rational approximation to the square root, $\sqrt{1-s^2}$, then (2.4) is replaced by

$$\xi = \frac{\omega r(s)}{c} \tag{2.5}$$

where p_m and q_n are polynomials of degree m and n in s respectively. Substituting in (2.5) for $p_m(s)$ and $q_n(s)$ and multiplying by the factor ω^K, $K = \max\{n, m-1\}$, to clear factors of ω in the denominator we obtain the dispersion relation for the differential equation

$$\sum_{j=0}^{n} q_j c^{j+1} u_{y^j\, t^{K-j}x} = \sum_{j=0}^{m} p_j c^j u_{y^j\, t^{K-j+1}}. \tag{2.6}$$

Trefethen and Halpern [12] proved that only rational approximations with $m = n$ or $m = n+2$ lead to well-posed problems. In [3] they employed a variety of techniques to derive well-posed approximations. In Table 2.1 we give the coefficients of these approximations for $k = 1, \ldots, 4$, $k = \frac{1}{2}(m+n+2)$. Observe that the $k = 2$ equation with $p_0 = 1$ and $p_2 = -1/2$ is the second order paraxial approximation of Clayton and Engquist [1]. Using the method of discretisation given by Clayton and Engquist for the paraxial equation we obtain the general form

$$D_+^x D_0^t u_{nk}^0 - \frac{p_0}{2c} D_+^t D_-^t (u_{nk}^0 + u_{nk}^1) - \frac{p_2 c}{2} D_+^y D_-^y (u_{n+1k}^1 + u_{n-1k}^0) = 0 \tag{2.7}$$

for the $k = 2$ equations

$$u_{xt} = \frac{p_0}{c} u_{tt} + c p_2 u_{yy}. \tag{2.8}$$

In (2.7) D_+^q, D_-^q and D_0^q are standard forward, backward and central difference operators, u_{nk}^j is an approximation to $u(j\Delta x, k\Delta y, n\Delta t)$, Δx, Δy are gridsizes in x and y directions

k	Padé	L_α^∞	Chebyshev points	L^2	C-P	Newman points	L^∞
1	1.00000	.99240	.70711	.78540	.63662	.00000	.50000
	0.00000	0.00000	0.00000	0.00000	0.00000	0.00000	0.00000
	0.00000	0.00000	0.00000	0.00000	0.00000	0.00000	0.00000
	0.00000	0.00000	0.00000	0.00000	0.00000	0.00000	0.00000
2	1.00000	1.00023	1.03597	1.03084	1.06103	1.00000	1.12500
	−.50000	−.51555	−.76537	−.73631	−.84883	−1.00000	−1.00000
	0.00000	0.00000	0.00000	0.00000	0.00000	0.00000	0.00000
	0.00000	0.00000	0.00000	0.00000	0.00000	0.00000	0.00000
3	1.00000	.99973	.99650	.99250	.99030	1.00000	.95651
	−.75000	−.80864	−.91296	−.92233	−.94314	−1.00000	−.94354
	0.00000	0.00000	0.00000	0.00000	0.00000	0.00000	0.00000
	−.25000	−.31657	−.47258	−.51084	−.55556	−.66976	−.70385
4	1.00000	1.00015	1.00034	1.00227	1.00161	1.00000	1.01773
	−1.00000	−1.16394	−1.27073	−1.37099	−1.37170	1.48698	−1.59044
	.12500	.22308	.29660	.38178	.38027	.48698	.57976
	−.50000	−.65974	−.76017	−.83407	−.84000	−.91384	−.94301

Table 2-1

Coefficients of one-way wave equations
Coefficients are listed in the pattern
$$p_0$$
$$p_2$$
$$\underline{p_4}$$
$$q_2$$

and Δt is the timestep. Renaut and Petersen [8] tested these equations numerically and noticed that the von Neumann stability bound, $\mu \leq \frac{1}{\sqrt{2}}$ for the wave equation with the standard five point stencil

$$u_{n+1k}^j - 2u_{nk}^j + u_{n-1k}^j = \mu^2 \{u_{nk}^{j+1} - 4u_{nk}^j + u_{nk}^{j-1} + u_{nk+1}^j + u_{nk-1}^j\}. \qquad (2.9)$$

was not achieved in all cases. Here $\mu = c\Delta t/\Delta x$ and we assume a square grid $\Delta x = \Delta y$. Subsequently they proved that the condition $\mu < -\frac{1}{2p_2}$ is necessary for the stability of (2.7) with (2.9). This confirmed their numerical results.

Our interest here is to avoid this restriction on μ by deriving alternative difference approximations. Now (2.7) has a truncation error $O(\Delta x^2) + O(\Delta t^2) + O(\Delta x \Delta t)$ because the first order terms cancel. By using Taylor Series expansions we can find other discretisations with the same truncation error for equation (2.7) and the higher order differential equations derived from the approximations of Table 2.1. Examples are given by Renaut [9]. This method is, however, tedious and has to be repeated for each value of k, and becomes more complicated as k increases. Instead we consider Lindman's formulation of his boundary condition [7].

Lindman [7] looked for an approximation to the inverse square root in equation

(2.4):

$$R(s) = \frac{A_m(s)}{D_n(s)} = \frac{1}{\sqrt{1-s^2}}. \tag{2.10}$$

For general approximations $R(s)$ we obtain a partial differential equation

$$\sum_{i=0}^{m} a_i c^{i+1} u_{y^i{}_t K-i_x} = \sum_{i=0}^{n} d_i c^i u_{y^i{}_t K-i+1} \tag{2.11}$$

where $K = \max\{m, n-1\}$ which is similar to equation (2.6). In [7] $R(s)$ is constrained to be of the form

$$R(s) = 1 + \sum_{i=1}^{N} \frac{\alpha_i s^2}{1 - \beta_i s^2} \tag{2.12}$$

which means that $a_0 = d_0 = 1$, and $D_n(s) = \Pi_{i=1}^{N}(1 - \beta_i s^2)$. The coefficients of the polynomials $A_m(s)$ and $D_n(s)$ are easily solved for if required. On the contrary, Lindman did not require these coefficients because he did not solve equation (2.11) directly. Rather, he substituted (2.12) into the dispersion relation (2.4) to give

$$i(\omega - \xi c) = i\left(\xi c \sum_{j=1}^{N} \frac{\alpha_j s^2}{1 - \beta_j s^2}\right) \tag{2.13}$$

which, upon associating $i\omega$, $i\xi$ and $i\eta$ with partial differentiation by t, x and y respectively gave the system of equations

$$u_t - cu_x = c\sum_{i=1}^{N} h_i$$
$$\frac{\partial^2 h_i}{\partial t^2} - \beta_i c^2 \frac{\partial^2 h_i}{\partial y^2} = \alpha_i c^2 \frac{\partial^2}{\partial y^2}\left(\frac{\partial u}{\partial x}\right). \quad i = 1, \ldots, N \tag{2.14}$$

Thus this system of $N + 1$ equations, $2N = \max\{n, m\}$ must be solved for at the boundary. The first of these is the normal incidence absorbing boundary condition modified by the correction functions h_i. These correction functions are the solutions of one-dimensional wave equations along the boundary.

In a similar way, by expressing the rational function $r(s)$ in the form

$$r(s) = p_0\left(1 + \sum_{i=1}^{N} \frac{\alpha_i s^2}{1 - \beta_i s^2}\right) \tag{2.15}$$

where $2N = \max\{m, n\}$, we obtain the system of equations

$$u_x - \frac{p_0}{c}u_t = \frac{p_0}{c}\sum_{i=1}^{N} h_i$$

where

$$\tag{2.16}$$

$$\frac{\partial^2 h_i}{\partial t^2} - \beta_i c^2 \frac{\partial^2 h_i}{\partial y^2} = c^2 \alpha_i \frac{\partial^3 u}{\partial y^2 \partial t} \quad i = 1, \ldots, N$$

that must be solved at the boundary instead of finding the solution to (2.6).

Equipped with this new formulation it is much easier to find a difference approximation at the boundary (2.16). Lindman solved the system (2.14) with the discrete approximation

$$D_+^t \frac{(u_{nk}^0 + u_{nk}^1)}{2} - cD_+^x \frac{(u_{n+1k}^0 + u_{nk}^0)}{2} = c\sum_{i=1}^{N}(h_i)_{nk}^0 \qquad (2.17)$$

and

$$(D_+^t D_-^t - \beta_i c^2 D_+^y D_-^y)(h_i)_{n-1k}^0 = \alpha_i \frac{c^2}{2}(D_+^y D_-^y)D_-^x(u_{nk}^0 + u_{n-1k}^0)$$

where $(h_i)_{nk}^0$ is an approximation to $h_i(0, k\Delta y, n\Delta t)$. The second order derivatives are approximated by a second order central difference and the first order derivatives by backward differences averaged over two levels. The equation for the h_i has a truncation error $O(\Delta x) + O(\Delta t)$. Renaut [9] showed that this does not degrade the results. Therefore we apply an equivalent difference scheme for equations (2.16):

$$cD_+^x(u_{n+1k}^0 + u_{nk}^0) - p_0 D_+^t(u_{nk}^0 + u_{nk}^1)$$

$$= p_0 \sum_{i=1}^{N}(h_i)_{nk}^0 \qquad (2.18)$$

$$(D_+^t D_-^t - \beta_i c^2 D_+^y D_-^y)(h_i)_{n-1k}^0$$

$$= c^2 \alpha_i D_+^y D_-^y D_+^t(u_{n-1k}^1 + u_{n-1k}^0).$$

The computational molecules for the $k = 2, 3$ and 4 equations given by (2.18) are illustrated in Figure 2.1. We observe that the molecule is different to that given by (2.7) for $k = 2$, as illustrated in Figure 2.2.

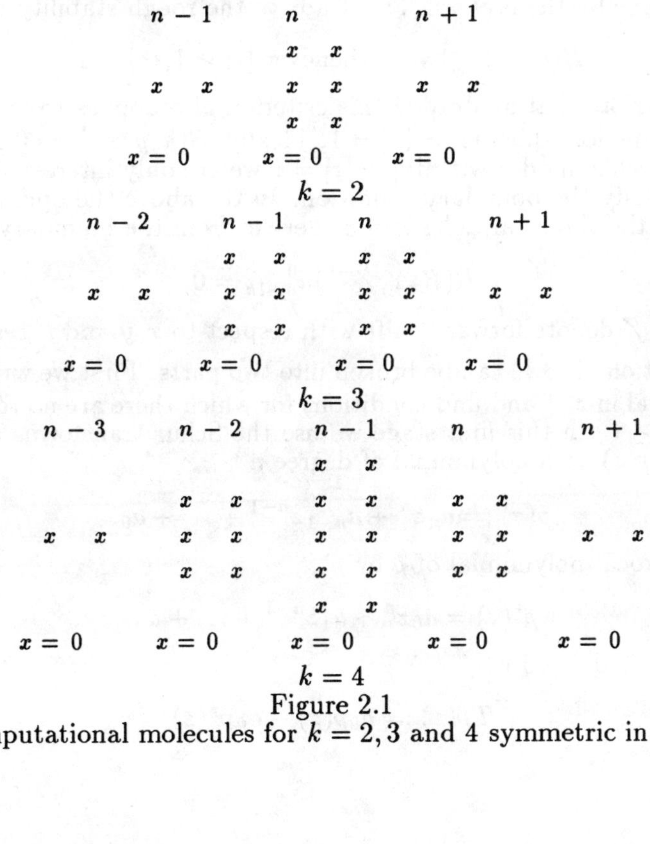

Figure 2.1
Computational molecules for $k = 2, 3$ and 4 symmetric in time.

$$n-1 \qquad n \qquad n+1$$

$$
\begin{array}{llllll}
x & & & & & x \\
x & x & & x \quad x & & x \quad x \\
x & & & & & x \\
x=0 & & x=0 & & x=0 &
\end{array}
$$

Figure 2.2
Computational molecule for $k=2$.
Clayton and Engquist form.

In all cases given by (2.18) these molecules are symmetric in time compared to (2.7) which is antisymmetric in time.

Note that these computational molecules are actually transparent when equations (2.18) are solved without expansion. However, for comparison, we eliminated the variables h in order to determine the dependencies. Furthermore, to study stability, the expanded form is required. Observe also that it is significantly simpler to program (2.15) for general k than it is to program each new boundary condition as k increases using the method discussed in Renaut [9]. Given the polynomial coefficients of $p_m(s)$ and $q_n(s)$ is it not difficult to find $\{\alpha_i, \beta_i, i = 1, \ldots, N\}$.

3. Stability. In the case of the solution of a problem with imposed boundary conditions the Gustafsson, Kreiss and Sundström (GKS) [2] theory demonstrates that it is not sufficient to merely apply a von Neumann stability analysis to the interior difference scheme. The physical interpretation of the GKS stability, as given by Higdon [5,6] and Trefethen [10,11], is that the boundary itself must not support exponentially growing waves nor should it support incoming waves that also satisfy the interior scheme by themselves. This leads to the rough stability criterion

$$B(\kappa, y, z^{-1}) \neq 0 \text{ whenever } |z| \geq 1, |\kappa| \leq 1 \tag{3.1}$$

of Higdon [5,6]. Note that as worded this criterion also applies to outgoing waves of the interior scheme for which $|z| = |\kappa| = 1$. Ideally, $B(\kappa, y, z^{-1}) = 0$ for these waves. Thus for the wavelike modes with $|\kappa| = |z| = 1$ we are only interested that incoming waves do not satisfy the boundary condition. In the above the operator B, which is a polynomial in the three variables κ, y, z, derives from the boundary condition

$$B(K, Y, Z^{-1})u_{n+1k}^0 = 0, \tag{3.2}$$

where K, Y and Z denote forward shift with respect to x, y and t, respectively.

The verification of (3.1) can be broken into two parts. First we write the operator B as a polynomial in z^{-1} and find conditions for which there are no roots that satisfy $|\kappa| < 1$ and $|z| \geq 1$. In this first stage we use the Schur transforms as described by Henrici [4]. Let $p(z)$ be a polynomial of degree n

$$p(z) := a_n z^n + a_{n-1} z^{n-1} + \ldots + a_0.$$

Define the reciprocal polynomial of p by p^*,

$$p^*(z) = \bar{a}_0 z^n + \bar{a}_1 z^{n-1} + \ldots + \bar{a}_n.$$

The polynomial T_p defined by

$$Tp(z) := \bar{a}_0 p(z) - a_n p^*(z)$$

is of degree $n - 1$ and is called the Schur transform of p. Furthermore,

$$Tp(0) = |a_0|^2 - |a_n|^2$$

is always real. We can repeat the process to find the Schur transform of Tp, T^2p. Note that the Schur transform of a polynomial of degree zero is the zero polynomial. Thus we have the iterated Schur transforms $T^k p$ defined by

$$T^k p := T(T^{k-1}p) \quad k = 2, 3, \ldots n$$

and we can set $\gamma_k := T^k p(0)$, $k = 1, 2, \ldots, n$. Then quoting from Henrici [4] we have conditions on the γ_k for the zeros of p to lie outside the unit circle:

Theorem 6.8 b Henrici [4]. *Let p be a polynomial of degree n, $p \neq 0$. All zeros of p lie outside the unit disk $|z| \leq 1$ if and only if $\gamma_k > 0$, $k = 1, 2, \ldots, n$.*

To apply this theory we consider B as a polynomial in $W = z^{-1}$ and find conditions under which the zeros satisfy $|W| > 1$, with $|\kappa| < 1$. As the order of the approximation to the square root increases we need to analyse $\gamma_k > 0$ for increasing k. For example, the order six approximations lead to a polynomial B which is quartic in N and hence $\gamma_1, \gamma_2, \gamma_3$ and γ_4 must be considered. The condition that $\gamma_i > 0$ imposes conditions on the Courant number μ in terms of the polynomial coefficients of the approximations. The results obtained for the Lindman type schemes are summarised in Table 2.2 and stated in the Theorem later in this section.

Of course, the above analysis is not complete as only part of condition (3.1) has been investigated. In the second stage we consider $|\kappa| = 1$. When $|\kappa| = 1$, however, there can be no solutions which satisfy both the interior five point scheme and the boundary condition with $|z| > 1$, see Higdon [6]. The roots with $|\kappa| = |z| = 1$ do, on the contrary, exist. For the five point scheme Higdon [6] has shown that the interior scheme supports waves pointing into the domain for which the arguments of κ and z are of different sign. If the boundary operator also supports these waves then waves which radiate spontaneously from the boundary and are transported by the interior can exist. These solutions are called generalised eigensolutions of the second kind. In Trefethen [11], where the first order hyperbolic equation is discussed, it is shown that, if these solutions exist, the instabilities that appear can be of different levels of severity, growing linearly with, or as the square root of, the number of time steps. In all cases reported there, the existence of an unbounded reflection coefficient compounds the problem. Furthermore, linear growth at the boundary can be converted to exponential by repeated reflection at a second boundary. Generalised eigensolutions of the first kind, those for which the group velocity is zero in the x-direction are in practice not so troublesome. These correspond to waves travelling up and down the boundary. Higdon [5] reported incompatibility between initial and boundary data as a mechanism which can excite instability in this zero frequency situation. This instability is, however, confined to the boundary.

In the stability theory modes of the form $\kappa^j e^{i\eta y} z^n$ arise from using a Fourier transform in the direction of the boundary and a Laplace transform in time. Hence, we consider only waves that are oscillatory in y and set $y = e^{i\eta y}$. In all cases we need to consider the roots for which $\eta \Delta y \neq 0$ and $\eta \Delta y = 0$ separately. The $\eta \Delta y = 0$ condition always leads to the zero frequency solution $(\kappa, z) = (1, 1)$. The case $\kappa = z = 1$, however, corresponds both to incoming and outgoing waves because $\kappa = 1$ is a double root of the operator equation for the five-point interior scheme. At the boundary either condition (3.1) is not satisfied $B(1, 1) = 0$, which means that both incoming and outgoing modes are annihilated or $B(1, 1) \neq 0$ and the reflection coefficient satisfies $|R| = 1$. In this latter case there is then total reflection at zero frequency and substantial reflection for low frequencies, $\Delta x, \Delta y \to 0$. It can be shown that $B(1,1) = 0$ for the examples here, and, thus, the zero frequency solutions are no problem.

For the roots with $|\kappa| = |z| = 1$ we found that for the $k = 2$ and $k = 4$ molecules we could solve for κ to get $\kappa = -\frac{\Omega}{\bar{\Omega}}$ when $z = e^{i\theta}$. Here $\Omega = \Omega(e^{i\theta})$. Thus $|\kappa| = 1$ and $\arg \kappa = 2 \arg \Omega + \pi$. For $\Omega = x + i\tilde{y} \sin \theta$, $\arg \Omega = \tan^{-1}\left(\frac{\tilde{y} \sin \theta}{x}\right)$ and thus if $y = \tilde{y} \sin \theta$ and x are of the same sign $\arg \kappa$ is negative. Therefore, if $\tilde{y}$ and x have the same signs the sign of $\arg \kappa$ is different to that of $\arg z$ and generalised eigensolutions may exist. For the $k = 3$ molecule given by (2.18), the analysis is similar. Since we can show that the roots with $|\kappa| = |z| = 1$ satisfy $\kappa = \frac{\Omega}{\bar{\Omega}}$, $\Omega = \tilde{x} \cos \theta/2 + i\tilde{y} \sin \theta/2$, generalised eigensolutions are possible when $\tilde{x}\tilde{y} < 0$. Furthermore, Renaut [9] has shown that where generalised eigensolutions exist $|R| < 1$ necessarily, where R is the reflection coefficient.

Renaut [9] has demonstrated analytically that, for all of the difference schemes considered here, generalised eigensolutions may exist. The analysis, however, is not conclusive and so numerical searches were carried out to determine when generalised eigensolutions exist. No modes satisfying both boundary and interior equations were found for the $k = 3$ case. For the $k = 2$ and $k = 4$ equations eigensolutions were found for $\mu = .1$ and $\mu = .3$ and the Newman boundary condtion. In these cases the reflection coefficient was also computed; not only did this confirm $|R| < 1$ but in fact $|R| << 1$ was found.

Thus, in summary, we present the conditions for stability for the $k = 2, 3, 4$ schemes obtained via (2.18). The assumptions about relations between the coefficients and their signs are satisfied by all the methods obtained from Table 2.1. Proofs of these results, which are long and detailed, are given in Renaut [9]. In each case, we assume that the interior scheme is solved by the five point difference scheme (2.9).

Theorem. *The difference schemes derived from equation 2.18 are stable only if*

(i) $k = 2$, $p_0 > 0$, $p_2 > 0$ and

$$\mu < \sqrt{\frac{p_0}{|p_2|}}, \quad \eta \Delta y \neq 0 \tag{3.3}$$

(ii) $k = 3$, $p_2 - p_0 q_2 < 0$, $p_0 > 0$, $q_2 < 0$, $p_2 > 0$, and

$$\mu < \min\left\{\sqrt{\frac{p_0}{|p_2|}}, \sqrt{\frac{p_0}{p_0 q_2 - p_1}}\right\} \tag{3.4}$$

(iii) $k = 4$, $p_0 q_2 - p_2 > 0$, $p_4 > 0$, $p_0 > 0$, $p_2 < 0$, $q_2 < 0$, $p_4 + q_2(p_0 q_2 - p_2) < 0$ and

$$\mu^2 < \min\left\{\frac{p_0}{p_0 q_2 - p_2}, \frac{-p_2 - \sqrt{p_2^2 - 4p_0 p_4}}{2p_4}, \right.$$
$$\left. \frac{(p_2 - p_0 q_2)p_0}{p_0 p_4 + p_2(p_0 q_2 - p_2)}\right\}, \eta \Delta y \neq 0 \tag{3.5}$$

We see in Table 3.1 that (3.3) is less restrictive than the bound $\mu < -\frac{1}{2}p_2$ for the Clayton and Engquist scheme (2.7). Additionally, (3.5) is less restrictive than the scheme derived by Renaut [9] for $k = 4$ whose computational molecule is given in Figure 3.1.

$n-3$	$n-2$	$n-1$	n	$n+1$
	x x		x x	
	x x	x x	x x	
x x	x x	x x	x x	x x
	x x	x x	x x	
	x x		x x	
$x=0$	$x=0$	$x=0$	$x=0$	$x=0$

Figure 3-1: computational molecule for $k=4$ scheme,
derived by Taylor series expansions

Numerical results of tests with these boundary conditions are presented by Renaut [9]. Except for the method represented by the computational molecule in Figure 3.1 the theoretically predicted bounds on μ are observed. This can be explained by the existence of generalised eigensolutions for this scheme for all values of μ. The instability is, however, only observed for small μ. For larger μ, the amount of reflected energy is larger than might be expected by comparison with the other $k=4$ scheme. The Newman boundary conditions are also not evidently unstable but exhibit greater reflected energy. These differences may be explained by the fact that where unstable behaviour is observed there are many more generalised eigensolutions. Where the instability only manifests itself with larger reflection we may suppose that the few eigensolutions existing do not have much energy and, thus, are not so troublesome.

We conclude that (2.18) is the appropriate finite difference scheme for the absorbing boundary conditions given here. Furthermore, the second and sixth order Newman boundary conditions should not be used because there is the possibility that the generalised eigensolutions are troublesome in different situations where, perhaps, the grid size is altered or a second boundary exists.

Approximation	Padé	L_α^∞	Chebyshev Points	L^2	C-P	Newman Points	L^∞
Method 2.7 $-\frac{1}{2}p_2$	1.0	0.9698	0.6533	0.6791	0.5890	0.5000	0.5000
Method 2.18 for $k=2$ $\sqrt{\frac{p_0}{|p_2|}}$	1.414	1.3929	1.1634	1.1832	1.1180	1.0000	1.0607
Method 2.18 for $k=3$ $\sqrt{\frac{p_0}{|p_2|}}$	1.1547	1.1119	1.0448	1.0373	1.0247	1.0000	1.0068
$\sqrt{\frac{p_0}{p_0 q_2 - p_2}}$	1.4142	1.4254	1.5041	1.5517	1.5952	1.7401	1.9234
Method of Fig. 3 $(\ \frac{1}{2}p_4(p_2+\sqrt{p_2^2+4p_0p_4})\ )^{1/2}$	.8990	.8667	.8243	.7893	.7891	.7530	.7308
$(\frac{1}{4}(p_2-p_0q_2+\sqrt{(p_0q_2-p_2)^2+8p_0p_4})\ )^{1/2}$	1.1118	1.0173	.9685	.9209	.9225	.8724	.8385
Method 2.18 for $k=4$ $\sqrt{\frac{p_0}{p_0 q_2 - p_2}}$	1.4142	1.4086	1.4001	1.3687	1.3743	1.3209	1.2643
$(\frac{-p_2-\sqrt{p_2^2-4p_0p_4}}{2p_4})^{1/2}$	1.0824	1.0416	1.0195	1.0108	1.0084	1.00	1.0012
$(\frac{(p_2-p_0q_2)p_0}{p_0p_4+p_2(p_0q_2-p_2)})^{1/2}$	1.1547	1.1775	1.2047	1.2363	1.2380	1.2526	1.2327

Table 3-1
Theoretical Bounds on Courant number μ.

4. Alternative Interior Schemes. Finally, we comment on the observation that results of our theorem are only valid with the discretisation (2.9). Our method of proof relies on showing that there can be no roots of the equation (3.2) for which $|\kappa| < 1$ and $|z| \geq 1$. These correspond to incoming waves that satisfy the equation (2.9) and that grow exponentially for $|z| > 1$, and are evanescent for $|z| = 1$. The oscillatory waves with $|\kappa| = |z| = 1$ are not allowed to satisfy the boundary condition with group velocity pointing into the domain. From (2.9) we get the condition that the arguments of κ and z must be of the same sign if such waves do not exist.

Suppose now that we consider a discretisation of the wave equation that is of higher order in space but still obtained using central differencing along the x- and y-directions. The weights of the points away from the centre of the stencil decrease in size rapidly and because we have used central differencing the method is symmetric in both directions. Therefore the roots κ of the operator form of the new method are symmetrically located relative to the unit circle for each z and, as for (2.9), the roots are reciprocals of each other. If we follow the analysis of Higdon [6] to determine the relationship between κ and z for incoming waves we see that the results will be the same, that $|\kappa| < 1$ for $|z| \geq 1$. This occurs because we analyse the oscillatory solutions and determine that the waves with incoming group velocity have arguments of κ and z that are of different sign. This is, of course, not the case for the general situation of arbitrary difference approximations to arbitrary hyperbolic equations but arises here due to the symmetry of the central difference approximations to the second derivatives in the wave equation. Then, when we consider the roots as we come off the unit circle and look specifically at roots for which $|\kappa|$ is near one the fact that the weights decrease in size away from the centre of the stencil means that the nearest points dominate and we can see that oscillatory solutions with incoming group velocity perturb into the unit circle for $|z| > 1$. Thus the results of Higdon follow and hence our stability bounds are appropriate in these cases too.

In section 3 we have derived conditions on the weights of the boundary operator in order that waves with $|\kappa| < 1$ and $|z| \geq 1$ do not satisfy the boundary condition. If, for any interior scheme, we can show that the only incoming waves satisfy $|\kappa| < 1$, $|z| > 1$, then we know that these are not supported by the boundary operators under the conditions given in section 3. As with the analysis for the symmetric interior operators analysis of generalized eigensolutions, $|\kappa| = |z| = 1$ must then be investigated separately.

An example shows that schemes that are central differenced but not in the usual way may also be covered by our theorem. In particular, we consider the five point stencil given by the rotation of the stencil (2.9) by $45°$ with h replaced by $\sqrt{2}h$. Here oscillatory solutions have group velocity pointing into the domain if the arguments of κ and z are of different sign and $\cos \eta \Delta y > 0$, otherwise the arguments of κ and z are of the same sign and $\cos \eta \Delta y < 0$. Thus, in this case, the analysis is performed for $\cos \eta \Delta y > 0$ and $\cos \eta \Delta y < 0$ separately. We soon see, however, that symmetry again imposes that incoming solutions satisfy $|\kappa| < 1$ for $|z| > 1$ and so the theorem is valid. The existence of generalised eigensolutions depends on the signs of $\cos \eta \Delta y$, $\arg \kappa$ and $\arg z$ and thus we need to reanalyse these solutions.

In any case it appears that the theorem is valid for any situation in which incoming solutions satisfy $|\kappa| < 1$, $|z| \geq 1$. The existence of generalised eigensolutions is harder to confirm and must probably be verified by performing a numerical search. Our results do demonstrate that, in the absence of generalised eigensolutions, the bounds on μ are valid since instability is observed for the $k = 2$ Clayton and Engquist scheme. Thus, the approach suggested here for investigating stability may be successfully applied for other interior schemes.

References

[1] R. CLAYTON and B. ENGQUIST, *Absorbing boundary conditions for acoustic and elastic wave equations*, Bull. Seis. Soc. Am., 67 (1977), pp. 1524–1540.

[2] B. GUSTAFSSON, H. O. KREISS and A. SUNDSTRÖM, *Stability theory of difference approximations for mixed initial boundary value problems* II, Math. Comp., 26 (1972), pp. 649–686.

[3] L. HALPERN and L. N. TREFETHEN, *Wide-angle one-way wave equations*, J. Accoust. Soc. Am., 84 (1988), pp. 1397–1404.

[4] P. HENRICI, *Applied and Computational Complex Analysis*, 1, John Wiley and Sons, Inc., (1974), pp. 491–496.

[5] R. L. HIGDON, *Absorbing boundary conditions for difference approximations to the multidimensional wave equation*, Math. Comp., 47 (1986), pp. 437–459.

[6] R.L. HIGDON, *Numerical absorbing boundary conditions for the wave equation*, Math. Comp., 49 (1987), pp. 65–90.

[7] E.L. LINDMAN, *Free space boundaries for the scalar wave equation*, J. Comp. Phys., 18 (1975), pp. 66–78.

[8] R. A. RENAUT and J. PETERSEN, *Stability of wide-angle absorbing boundary conditions for the wave equation*, Geophysics, 54 (1989), pp. 1153–1163.

[9] R. A. RENAUT, *Absorbing boundary conditions, difference operators and stability*, submitted J. Comp. Phys, 1989.

[10] L. N. TREFETHEN, *Group velocity interpretation of the stability theory of Gustafsson, Kreiss and Sundström*, J. Comp. Phys., 49 (1983), pp. 199–217.

[11] L. N. TREFETHEN, *Instability of difference models for hyperbolic initial boundary value problems*, Comm. Pure. Appl. Math., 37 (1984), pp. 329–367.

[12] L. N. TREFETHEN and L. HALPERN, *Well-posedness of one-way wave equations and absorbing boundary conditions*, Math. Comp., 47 (1986), pp. 421–435.

CHAPTER 8

'Optimal Filtering' for Some Ill-Posed Problems[1]

Thomas I. Seidman*

ABSTRACT: We describe an approach to a class of ill-posed problems in which the determination of a 'filter' for obtaining approximate solutions is obtained by means of an optimization process. In Hilbert space settings a fairly explicit computation may be possible and this is presented. It is noted that, under certain the resulting filter is, indeed, optimal in the sense of realizing a minimal uniform error bound.

1.   INTRODUCTION

An approach, which we have called 'optimal filtering', was introduced in [3] for the determination of approximate solutions for the 'sideways heat equation'. The approach provided a 'filter' in two senses: (a) its implementation took the form of a multiplication operator on a Fourier transform of the data and (b) its purpose was to extract as much useful information as possible while rejecting 'noise' in the data. The name then came both from the use of an optimization procedure for the determination of the appropriate filter to produce the approximate solution as well as from the fact that this particular filtering procedure did result in an optimal error bound.

The context for that approach involved a semigroup of 'smoothing' operators whose approximate inversion was the assigned task and it is anticipated that a basic context for generalizations of this approach would be a *parametrized family* of operators for which one wished to 'march backward'

* Department of Mathematics and Statistics, University of Maryland Baltimore County, Baltimore, MD 21228, USA.
[1] This research has been partially supported by the U.S. Air Force Office of Scientific Research under the grant AFOSR-87-0190 and by the National Science Foundation under the grant ECS-8814788.

with respect to the parameter. Numerous applications having this general structure come to mind. Here, in order to present the method in the clearest fashion, we consider the approach only as applied in the context of a *single* ill-posed problem although some comments will be made about the application to families of problems; see Remarks 4.4, 4.5. Also, in order to present the approach most clearly as 'a thing in itself' we treat the method abstractly here and make no attempt to work out any specific application, geophysical or otherwise.

Thus, let $x \in \mathcal{X}$ be an 'unknown of interest' and suppose we are given the following two pieces of information:

$$(1.1) \qquad .1) \qquad x = \mathbf{B}z \text{ with } \|z - \varphi_0\| \leq \varepsilon_0$$
$$.2) \qquad \|\mathbf{A}x - \varphi_1\| \leq \varepsilon_1$$

which we assume are consistent. Our primary attention is devoted to the case of continuous *linear* operators $\mathbf{A} : \mathcal{X} \to \mathcal{Y}$ and $\mathbf{B} : \mathcal{Z} \to \mathcal{X}$ with $\mathcal{X}, \mathcal{Y}, \mathcal{Z}$ uniformly convex Banach spaces, usually a Hilbert space $\mathcal{H}$ — although we note that some of our considerations also apply, more generally, to nonlinear maps and/or more general spaces. It is convenient to set

$$(1.2) \quad \mathcal{U}_0 = \mathcal{U}_0[\varphi_0, \varepsilon_0] \ := \ \{x \in \mathcal{X} : x = \mathbf{B}z \text{ with } \|z - \varphi_0\| \leq \varepsilon_0\},$$
$$\mathcal{U}_1 = \mathcal{U}_1[\varphi_1, \varepsilon_1] \ := \ \{x \in \mathcal{X} : \|\mathbf{A}x - \varphi_1\| \leq \varepsilon_1\}$$

so the set of solutions of (1.1) is

$$\mathcal{U} = \mathcal{U}[\varphi_0, \varepsilon_0; \varphi_1, \varepsilon_1] := \mathcal{U}_0[\varphi_0, \varepsilon_0] \cap \mathcal{U}_1[\varphi_1, \varepsilon_1].$$

Consistency is just the requirement that $\mathcal{U}$ be nonempty. If we assume that $\mathbf{A}$ is injective, then exact knowledge of $\bar{\varphi} := \mathbf{A}\bar{x}$ would uniquely determine $\bar{x}$ from $(1.1.2)$ with $\varepsilon_1 = 0$ even without $(1.1.1)$. On the other hand, we will assume that $\mathbf{A}$ does not have a *continuous* inverse so the problem is ill-posed and the uncertainty inherent in $(1.1.2)$ as stated would be catastrophically amplified by any attempt at (approximately) inverting $\mathbf{A}$ without some suitable stabilization — here to be provided by $(1.1.1)$. The stabilizing effect of such a condition had been noted long ago, cf., e.g., [5].

For our interpretation, we view the first condition $(1.1.1)$ as *a priori* information and view $(1.1.2)$ as corresponding to obtaining a value φ_1 by a 'noisy measurement', characterized by the operator $\mathbf{A}$ and the accuracy bound ε_1. Indeed, a standard form of analysis would be to assume existence of a 'true' solution $x = \bar{x}$ satisfying $(1.1.1)$ and giving $\mathbf{A}\bar{x} = \bar{\varphi}_1$ (with $\bar{\varphi}_1$ 'unknown'). There is then no suggestion that the 'regularizing bound' ε_0 should be 'small' but with $\varepsilon_1 \to 0$, one imagines a (potential) *sequence of increasingly accurate measurements:* $\varphi_1 = \varphi_1(\varepsilon_1)$ satisfying $\|\varphi_1 - \bar{\varphi}\| \leq \varepsilon_1$. One then seeks a way of using (1.1) to select— presumably by a computationally implementable procedure from knowledge of $\mathbf{A}, \mathbf{B}$ and the available data quadruple $[\varphi_0, \varepsilon_0; \varphi_1, \varepsilon_1]$ — some specific 'approximate solution' $v = v(\varepsilon_1)$ for which one would then wish to demonstrate that $v \to \bar{x}$ (as rapidly as possible) as $\varepsilon_1 \to 0$. Compare [12]. This procedure is a form of 'filtering', eliminating the effect of the noise, which is complicated in our present analysis by the ill-posedness-induced error amplification. To the extent that it is reasonable to consider $\varepsilon_0, \varepsilon_1$ not merely as bounds but as fairly sharp estimates of the 'signal strength' $\|z\|$ and of the 'noise' (measurement error), respectively, we may interpret the scaled parameter

$$\nu := \varepsilon_1/\varepsilon_0$$

as a *noise–to–signal ratio*.

In this setting the *data* for the problem, apart from the specification of the operators involved, consists of the quadruple $[\varphi_0, \varepsilon_0; \varphi_1, \varepsilon_1]$. While one could ask to know the *set* of solutions $\mathcal{U}$, one normally would only seek to specify a pair $[v, \varepsilon]$ such that

(1.3) $$\|x - v\| \leq \varepsilon \text{ for all } x \in \mathcal{U} = \mathcal{U}[\varphi_0, \varepsilon_0; \varphi_1, \varepsilon_1]$$

so ε is an error bound for the 'approximate solution' v of (this particular instance of) (1.1). We would, of course, like ε to be as small as possible and, to the extent that this is the case, will refer to an approximation procedure for computing the pair $[v, \varepsilon]$ as 'optimal'.

Rather than considering an approximate solution v for (1.1) in isolation for a specific data pair $[\varphi_0, \varphi_1]$, we seek a map

$$\tilde{\mathbf{F}} = \tilde{\mathbf{F}}_{[\varepsilon_0, \varepsilon_1]} : \mathcal{Z} \times \mathcal{Y} \to \mathcal{X} : [\varphi_0, \varphi_1] \mapsto v$$

giving an approximate solution $v = \tilde{\mathbf{F}}[\varphi_0, \varphi_1]$ for *each* data pair. We then refer to $\bar{\varepsilon}$ as a (realizable) **uniform bound** and to the map $\tilde{\mathbf{F}}$ as a **filter** realizing this bound if (1.3) holds with $\varepsilon = \bar{\varepsilon}$ for all instances of (1.1), i.e., if for all[2] $[\varphi_0, \varphi_1]$ one has

(1.4) $$\|x - \tilde{\mathbf{F}}[\varphi_0, \varphi_1]\| \leq \bar{\varepsilon} \quad \text{for all } x \text{ satisfying (1.1).}$$

If this $\bar{\varepsilon}$ is a **minimal uniform bound (MUB)** — i.e., if the infimum of all such realizable uniform bounds is itself realizable — then any filter $\tilde{\mathbf{F}}$ which does, indeed, realize $\bar{\varepsilon}$, giving (1.4), will be called an **optimal filter** for (1.1).

The operator $\mathbf{B}$ of the 'regularizing condition' (1.1.1) is typically linear and, in this case, we note a simplifying reduction of the problem: set

(1.5) $$\tilde{\varphi} := (\varphi_1 - \mathbf{AB}\varphi_0)/\varepsilon_0$$
$$\tilde{\mathbf{A}} : \mathcal{X} \to \mathcal{Y} \quad : \quad \tilde{x} \mapsto [\mathbf{A}(\mathbf{B}\varphi_0 + \varepsilon_0 \tilde{x}) - \mathbf{AB}\varphi_0]/\varepsilon_0$$

and consider the reduced problem

---

[2] While this statement reads 'for *all*', it is clear that the implication is nonempty only subject to consistency.

(1.6)   .1) $\tilde{x} = \mathbf{B}\tilde{z}$ with $\tilde{z} \in \mathcal{B} := \{z \in \mathcal{Z} : \|z\| \leq 1\}$

             .2) $\|\tilde{\mathbf{A}}\tilde{x} - \tilde{\varphi}\| \leq \nu$

with solution set

$$\tilde{\mathcal{U}}_\nu(\tilde{\varphi}) := \{\tilde{x} \in \mathbf{B}\mathcal{B} : \|\tilde{\mathbf{A}}\tilde{x} - \tilde{\varphi}\| \leq \nu\}.$$

A simple computation shows the effective equivalence of (1.6) to (1.1) in the sense that x is a solution of (1.1) if and only if $\tilde{x} := (x - \mathbf{B}\varphi_0)/\varepsilon_0$ is a solution of (1.6) with the 'target' and operator modified as in (1.5) — i.e., one has

$$(1.7) \qquad \tilde{\mathcal{U}}_\nu(\tilde{\varphi}) = \frac{1}{\varepsilon_0}\left(\mathbf{B}\varphi_0 + \mathcal{U}[\varphi_0, \varepsilon_0; \varphi_1, \varepsilon_1]\right).$$

In the context of this reduced problem (1.6) we call $\varepsilon > 0$ a (realizable) *uniform bound* if there is a *filter*, i.e., a map $\mathbf{F} = \mathbf{F}_\nu : \mathcal{Y} \to \mathcal{X} : \tilde{\varphi} \mapsto \tilde{v}$ such that

$$(1.8) \qquad \left[x \in \tilde{U}_\nu(\tilde{\varphi}) \Rightarrow \|x - \mathbf{F}_\nu\tilde{\varphi}\| \leq \epsilon\right] \qquad \text{for all } \tilde{\varphi} \in \mathcal{Y}.$$

We will denote by $\varepsilon_\# = \varepsilon_\#(\nu)$ a **minimal uniform bound** in this sense (with corresponding **optimal filter** $\mathbf{F}_\nu$). We state the following lemma; proof obvious.

LEMMA 1.1:   *A filter $\mathbf{F}_\nu$ realizes a bound ϵ for (1.6) as in (1.8) if and only if — for arbitrary $\varepsilon_0, \varepsilon_1$ giving $\nu = \varepsilon_1/\varepsilon_0$ and using (1.5) — the filter*

$$\tilde{\mathbf{F}} = \tilde{\mathbf{F}}_{[\varepsilon_0, \varepsilon_1]} : [\varphi_0, \varphi_1] \mapsto \mathbf{F}_\nu\tilde{\varphi}$$

realizes the uniform bound $\varepsilon_0\epsilon$ for (1.1) as in (1.3). Thus, $\bar{\varepsilon}$ is a MUB for (1.1) if and only if $\bar{\varepsilon} = \varepsilon_0\varepsilon_\#(\nu)$ with $\varepsilon_\#(\nu)$ a MUB for (1.6); further, the notions of optimal filter correspond as noted. ∎

2.   GENERAL RESULTS

For the problem (1.6) our task will be the development of an optimal (or suboptimal) filter $\mathbf{F}_\nu : \mathcal{Y} \to \mathcal{X}$ satisfying (1.8) and the principal result of this paper, to be presented in the next section, is a computation in the special case of commuting normal operators on a Hilbert space. We begin, however, by noting some results of a less explicit character but in more general settings.

THEOREM 2.1:   *Suppose $\mathcal{X}$ is uniformly convex and $\mathcal{U}$ is a bounded set in $\mathcal{X}$. Then there is a unique pair $[v_*, \varepsilon_*]$ minimizing $\varepsilon = \varepsilon_*$ subject to (1.3); we have $v_* \in \overline{co}\,\mathcal{U}_0$, the closed convex hull of $\mathcal{U}_0$.*

Further, if $\mathcal{U}$ has the form[3] $\{x \in \mathcal{U}_0 : \|\mathbf{A}x - \varphi_1\| < \varepsilon_1\}$ with $\mathcal{U}_0$ precompact and $\mathbf{A}$ continuous, then this $[v_*, \varepsilon_*]$ depends continuously on the pair $[\varphi_1, \varepsilon_1]$ with $\varepsilon_1 > 0$.

PROOF: Consider a sequence $\{[v_k, \varepsilon_k]\}$, each satisfying (1.3), for which we may assume $\varepsilon_k \searrow \varepsilon_* := \inf$ and (possibly extracting a subsequence by weak compactness, as $\{v_k\}$ is necessarily bounded) $v_k \rightharpoonup v_*$. As the norm is weakly lsc, this $[v_*, \varepsilon_*]$ satisfies (1.3) so the minimum is attained. One has uniqueness of v_* as otherwise $\mathcal{U}$ would be contained in the intersection of two balls with the same minimal radius ε_* but distinct centers, hence, by uniform convexity, in a strictly smaller ball with center at their average — which contradicts minimality. Similarly, if we did not have $v_* \in \overline{co}\,\mathcal{U}$ then replacing v_* by the nearest point in $\overline{co}\,\mathcal{U}$ would contradict minimality. (Cf., [6].)

Now supposing, as above, that $\mathcal{U} \subset \mathcal{U}_0$ with $\mathcal{U}_0$ precompact, we $\overline{co}\,\mathcal{U}$ compact. Consider a sequence $[\varphi_1^{[k]}, \varepsilon_1^{[k]}] \to [\varphi_1^*, \varepsilon_1^*]$ and the corresponding $[v_*^{[k]}, \varepsilon_*^{[k]}]$. Extracting a subsequence if necessary, we may assume that $v_*^{[k]} \to \bar{v}_*$ and $\varepsilon_*^{[k]} \to \bar{\varepsilon}_*$. If $x \in \mathcal{U}[\varphi_1^*, \varepsilon_1^*]$ then also $x \in \mathcal{U}[\varphi_1^{[k]}, \varepsilon_1^{[k]}]$ for large enough k so $\|x - v_*^{[k]}\| \le \varepsilon_*^{[k]}$. Going to the limit gives (1.3) and a similar argument shows minimality. The uniqueness above shows that this is independent of the extraction of a subsequence.   ■

It is clear that the definition above of $[v_*, \varepsilon_*]$ gives the best information about 'the solution' of any particular instance of (1.1) which could possibly be provided in such a form — an approximant with a minimal error bound — and an algorithm which would determine v_* would certainly be truly optimal. For our present analysis, however, we are content to seek an optimal filter in the sense described above — for which it is the *uniform* bound which is to be minimal.

THEOREM 2.2:   *Let the Banach space $\mathcal{X}$ be reflexive and $\mathbf{B}$ linear. Then for every $\varepsilon_0, \varepsilon_1 > 0$ there is always an optimal filter $\tilde{\mathbf{F}} = \tilde{\mathbf{F}}_{[\varepsilon_0, \varepsilon_1]}$ realizing a minimal uniform bound $\bar{\varepsilon}$ as in (1.4).*

PROOF: We actually prove the result somewhat more generally than stated — we need not assume any linearity of $\mathbf{A}, \mathbf{B}$ but only that $\mathcal{X}$ is

_____

[3]Observe that this does not quite fit the framework of (1.1.2) in that it effectively replaces "$\le \varepsilon_1$" by "$< \varepsilon_1$". Using nonlinear $\mathbf{A}$, one can easily construct counterexamples to the continuity without this modification although the modification seems irrelevant in the linear case. We also note that if $\mathbf{A}$ is injective then continuity in φ_1 for the 'unmodified' case with $\varepsilon_1 = 0$ is a classical theorem of topology.

reflexive and that (for given $\varepsilon_0, \varepsilon_1 > 0$) the solution sets $\mathcal{U}[\varphi_0, \varepsilon_0; \varphi_1, \varepsilon_1]$ all have finite diameter, bounded uniformly as $[\varphi_0, \varphi_1]$ varies over $\mathcal{Z} \times \mathcal{Y}$. This is already the case if $\mathbf{B}$ is uniformly Lipschitzian, much less linear.

Under these circumstances realizable uniform bounds exist since one can obviously always select a center v for a ball large enough to contain each solution set; such selections define a 'filter'. Now let $\bar{\varepsilon}_k \searrow \bar{\varepsilon}$ be a minimizing sequence with corresponding filters $\mathbf{F}_k$. For any particular instance of (1.1), set $v_k := \mathbf{F}_k[\varphi_0, \varphi_1]$. Note that (unless $\mathcal{U}$ is empty so one could arbitrarily set $v_k = 0$) the sequence $\{v_k\}$ must be bounded. Extracting a subsequence if necessary, we can assume $v_k \rightharpoonup v =: \mathbf{F}[\varphi_0, \varphi_1]$. Having (1.3) for each $[v_k, \bar{\varepsilon}_k]$, the lower semicontinuity of the norm with respect to the weak topology now ensures (1.3) for $[v, \bar{\varepsilon}]$. Doing this for each pair $[\varphi_0, \varphi_1]$ realizes the minimal $\bar{\varepsilon}$ by the optimal filter $\mathbf{F}$. ∎

THEOREM 2.3:   *Let $\mathcal{X}, \mathcal{Y}, \mathcal{Z}$ be uniformly convex Banach spaces; let $\mathbf{A} : \mathcal{X} \to \mathcal{Y}$ and $\mathbf{B} : \mathcal{Z} \to \mathcal{X}$ be continuous linear operators with $\mathbf{B}$ compact and $\mathbf{A}$ injective. Then the minimal uniform bound $\varepsilon_{\#}(\nu)$ for the reduced problem (1.6) satisfies $e_{\#}(\nu) \to 0$ as $\nu \to 0$.*

PROOF: We will use a lemma from [11]:[4]

LEMMA: *Let $\mathcal{X}, \mathcal{Y}, \mathcal{Z}$ be Banach spaces with $\mathcal{Z}$ reflexive; let $\mathbf{B} : \mathcal{Z} \to \mathcal{X}, \mathbf{C} : \mathcal{Z} \to \mathcal{Y}$ be continuous linear operators with $\mathbf{B}$ compact and $\mathcal{N}(\mathbf{C}) \subset \mathcal{N}(\mathbf{B})$. Then, for any $\delta > 0$ there exists C_δ such that, with the appropriate norms,*

$$\|\mathbf{B}z\| \leq \delta\|z\| + C_\delta\|\mathbf{C}z\|$$

for all $z \in \mathcal{Z}$.

PROOF: Suppose not. Then there is a sequence $z_n \in \mathcal{Z}$ with $\|z_n\| = 1$, so we may assume $z_n \rightharpoonup \tilde{z}$, such that $\|\mathbf{B}z_n\| > \delta + n\|\mathbf{C}z_n\|$. Then $\mathbf{C}z_n \to 0$ and $\mathbf{C}\tilde{z} = 0$ whence, as $\mathcal{N}(\mathbf{C}) \subset \mathcal{N}(\mathbf{B})$, one has $\mathbf{B}\tilde{z} = 0$. As $\mathbf{B}$ is compact we have strong convergence $\mathbf{B}z_n \to \mathbf{B}\tilde{z}$ which, as $\|\mathbf{B}z_n\| > \delta$, gives $\|\mathbf{B}\tilde{z}\| \geq \delta > 0$ — a contradiction. ∎

Now let $\mathcal{K}_0 := \{\mathbf{A}x : x \in \mathcal{U}_0\}, \mathcal{K} := \{\mathbf{A}x : x \in \overline{\mathcal{U}_0}\}$; as $\mathbf{A}$ is continuous, $\mathcal{K}$ is closed. Setting $\mathcal{B}_1 := \{y \in \mathcal{Y} : \|y\| \leq \varepsilon_1\}$, let $\mathcal{V}_1 = \mathcal{V}_1(\varphi_1) := \mathcal{K} \cap [\varphi_1 - \mathcal{B}_1]$ so $\mathcal{U} = \{x \in \mathcal{U}_0 : \mathbf{A}x \in \mathcal{V}_1\}$. Note that $\mathcal{K}$ is convex so, given φ_1 outside $\mathcal{K}$, we may find a 'nearest point' $\tilde{\varphi}_1$ in $\mathcal{K}$ and have $\mathcal{V}_1 \subset \tilde{\mathcal{V}}_1 := \mathcal{V}_1(\tilde{\varphi}_1)$; thus we need only consider $\varphi_1 \in \mathcal{K}$.

For any given $\varphi_1 \in \mathcal{K}$, the density of $\mathcal{K}_0$ enables us to find $\bar{\varphi}_1 \in \mathcal{K}_0$ with $\|\varphi_1 - \bar{\varphi}_1\| \leq r$. By definition we have $\bar{\varphi}_1 = \mathbf{C}\tilde{z}$ for some $\tilde{z} \in \mathcal{B} := \{z \in \mathcal{Z} : \|z\| \leq 1\}$ with $\mathbf{C} := \mathbf{A}\mathbf{B}$ and we set $v := \mathbf{B}\tilde{z}$. Note that this necessarily

---

[4]Note, also, the quite similar lemma in [13].

gives v in the compact set $\overline{\mathcal{U}_0}$. For any $x \in \mathcal{U}$, we have $x = \mathbf{B}z$ with $z \in \mathcal{B}$ (so $\|z - \tilde{z}\| \leq 2$) and note that $x - v = \mathbf{B}(z - \tilde{x})$ and

$$\|\mathbf{C}(z - \tilde{z})\| \leq \|\mathbf{A}x - \varphi_1\| + \|\varphi_1 - \tilde{\varphi}_1\| \leq 2r.$$

We may then apply the Lemma to obtain

$$(2.1) \qquad\qquad \|x - v\| \leq \varepsilon := 2\delta\varepsilon_0 + 2rC_\delta$$

(with $\delta > 0$ arbitrary) since compactness of $\mathbf{B}$ makes $\mathbf{C} = \mathbf{AB}$ compact and injectivity of $\mathbf{A}$ gives $\mathcal{N}(\mathbf{C}) = \mathcal{N}(\mathbf{B})$. The estimate (2.1) gives (1.3) with ε uniform for all φ_0, φ_1 once one chooses δ. This bound ε need not be minimal but if (1.3) is satisfied for a sequence $[v_k, \varepsilon_k]$ with $\{\varepsilon_k\}$ a minimizing sequence and $v_k \in \overline{\mathcal{U}_0}$ as above, then we may extract a subsequence giving $v_k \to \tilde{v}$ and $\varepsilon_k \to \varepsilon_\# := \inf$ and have (1.3) satisfied using $[\tilde{v}, \varepsilon_\#]$. Since this $\varepsilon_\#$ is bounded by the $\varepsilon = \varepsilon_\delta$ presented in (2.1), we note that that can arbitrarily small by choosing δ small and then, having fixed C_δ, considering small enough r. ∎

3.   PRINCIPAL RESULTS

At this point we turn to the Hilbert space computation which is our principal result. We will assume that $\mathbf{A}, \mathbf{B}$ are commuting normal operators on a Hilbert space $\mathcal{H}$ ($\mathcal{X}, \mathcal{Y}, \mathcal{Z} = \mathcal{H}$) and that $\mathbf{A}$ is injective. We are, of course, primarily concerned with situations in which $\mathbf{A}$ is not boundedly invertible, i.e., 0 is in the spectrum $\sigma(\mathbf{AB})$ but is not an eigenvalue of $\mathbf{A}$.

Before even stating the theorem we note that in treating (1.6) it is always irrelevant to consider any $\nu > \bar{\nu} := \|\mathbf{AB}\|$ since one could always then replace the data pair $[\tilde{\varphi}, \nu]$ in (1.6.2) by $[0, \bar{\nu}]$ on the basis of (1.6.1) alone. It will also be convenient to recall a standard[5] spectral theory result:

> One may, to within a unitary isomorphism, take $\mathcal{H}$ to be $L^2_\mu(\Omega)$ for some measure space[6] (Ω, Σ, μ) in such a way that $\mathbf{A}, \mathbf{B}$ simultaneously become multiplication operators:
>
> $$(3.1) \qquad \mathbf{A}x = \alpha(\cdot)x(\cdot), \qquad \mathbf{B}x = \beta(\cdot)x(\cdot)$$
>
> for $x = x(\cdot) \in \mathcal{H} = L^2_\mu(\Omega)$.

This is somewhat more convenient than spectral measures or eigenfunction expansions in permitting us to work with the pair of operators. For a variety

---

[5] For this see, e.g., Theorem 9.1 of [10] which gives this for the self-adjoint case only. By looking separately at real and imaginary parts and noting Theorem 12.16 of [9]), one sees that it can also be applied to commuting normal operators as here.

[6] If either $\mathbf{A}$ or $\mathbf{B}$ is compact, then this is necessarily discrete and $\mathcal{H}$ is effectively the sequence space ℓ^2.

of (constant coefficient) differential operators, of course, this becomes just the usual 'frequency domain': the spectral resolution (3.1) is obtained by taking $\omega \in \Omega$ to be the 'Fourier transform variable'; the typical 'engineering' notion of a filter would be a multiplication operator in this representation. In any case, we shall here assume that this resolution is effectively available.

The key to our approach here is a function $\psi : \Sigma^+ \to [0, \|\mathbf{B}\|]$ where

$$\Sigma^+ := \operatorname{ess\,ran}(|\alpha\beta|) = \{t = |\gamma| : \gamma \in \sigma(\mathbf{AB}) = \operatorname{ess\,ran}(\alpha\beta)\}$$

measuring the size of $|\beta|$ in terms of $|\alpha\beta|$. We want, sharply,

$$(3.2) \qquad |\beta| \le \psi(|\alpha\beta|) \qquad \text{ae on } \Omega;$$

roughly, we want $\psi(t) = \sup\{|\beta| : |\alpha\beta| = t\}$. This would, indeed, be an adequate definition for 'nice' α, β but is technically inadequate for merely measurable α, β. What we need, besides (3.2), is that, for any $t \in \Sigma^+$, there should be a set $\mathcal{S}_t$ of positive measure in Ω, on which one has α, β simultaneously (approximately) constant: $|\alpha\beta| \approx t$ with $|\beta| \approx \psi(t)$ (so $|\alpha| \approx t/\psi(t)$). Thus we define

$$(3.3) \quad \psi(t; \delta) := \operatorname*{ess\,sup}_{\omega}\{|\beta| : ||\alpha\beta| - t| < \delta\}, \quad \psi(t) := \lim_{\delta \searrow 0} \psi(t; \delta).$$

Note that the set $\{\omega : ||\alpha\beta| - t| < \delta\}$ has positive measure for each $\delta > 0$ precisely when $t \in \Sigma^+$. It is clear that we then have (3.2) and that there is necessarily a set $\mathcal{S}_{t;\delta}$ of positive measure such that

$$(3.4) \qquad ||\alpha\beta| - t| < \delta \text{ and } ||\beta| - \psi(t)| < \delta \text{ on } \mathcal{S}_{t;\delta}.$$

In terms of these definitions we can state our result.

THEOREM 3.1: *Let* $\mathbf{A}, \mathbf{B}$, *etc., be as above. Then (for the reduced problem as in Lemma 1.1 for any* $\nu > 0$*) the minimal uniform bound* $\varepsilon_\#(\nu)$ *satisfies*
$$(3.5) \qquad \hat{\varepsilon}_-(\nu) \le \varepsilon_\#(\nu) \le \hat{\varepsilon}_+(\nu) \qquad \text{where}$$

$$(3.6) \qquad \begin{cases} \hat{\varepsilon}_-(\nu) := & \sup\{\psi(t) : \nu \ge t \in \Sigma^+\} \\ \\ \hat{\varepsilon}_+(\nu) := & \min_{s>0}\{\nu s + \sup_{t \in \Sigma^+}\{\psi(t) - st\}\}. \end{cases}$$

The uniform bound $\hat{\varepsilon}_+(\nu)$ *is realizable by a linear filter* $\mathbf{F}$ *(also commuting with* $\mathbf{A}, \mathbf{B}$*), explicitly computable by an optimization.*[7] *Further, assuming* $\Sigma^+ = \operatorname{ess\,ran}(|\alpha\beta|)$ *is an interval, one can improve (3.5) when* ψ *is an increasing concave function: one then has*

---

[7]In some sense, it is just the use of such optimization to define this linear filter which justifies our use of the term 'method of optimal filtering' — as well, of course, as the possibility that this may be an optimal filter.

(3.7) $\hat{\varepsilon}_-(\nu) = \varepsilon_\#(\nu) = \hat{\varepsilon}_+(\nu) = \psi(\nu)$

so the filter defined here is an optimal filter.

PROOF: We wish to construct a linear filter for the reduced problem as a multiplication operator:

(3.8) $\mathbf{F} = \mathbf{F}_\nu : \varphi_1 \mapsto v(\cdot) := \vartheta(\cdot)\varphi_1(\cdot)$

for a suitably chosen 'filter factor' $\vartheta = \vartheta_\nu(\cdot)$. We begin with the observation that any solution of the problem has the form $x = \beta z$ with $z \in \mathcal{B} := \{y \in \mathcal{H} : \|y\| \leq 1\}$ by (1.6.1) and

(3.9)   error $:= x - \mathbf{F}\varphi_1 = (1 - \vartheta\alpha)\beta z + \nu\vartheta\zeta$ with $\zeta := [\alpha\beta z - \varphi_1]/\nu$

where $\zeta \in \mathcal{B}$ by (1.6.2).

We will explicitly construct ϑ for (3.8) so as to realize the bound $\hat{\varepsilon}_+(\nu)$ of (3.6). From (3.8), (3.9) we have

$$\begin{aligned} \|x - \mathbf{F}\varphi_1\| &\leq \|(1 - \vartheta\alpha)\beta z\| + \nu\|\vartheta\zeta\| \\ &\leq \sup_\Omega\{|1 - \vartheta\alpha||\beta|\} + \nu\sup_\Omega\{|\vartheta|\} \end{aligned}$$

(3.10)

as we have $\|z\|, \|\zeta\| \leq 1$. This suggests the key step of the method:

(3.11)   **optimization problem:**
   1.   given $s > 0$, choose $\vartheta(\omega)$ to minimize $|1 - \vartheta\alpha||\beta|$
            over $\vartheta \in \mathbb{C}_s := \{\vartheta \in \mathbb{C} : |\vartheta| \leq s\}$;
   2.   choose s to minimize $\sup_\Omega$ of the result.

For the first step, choosing each $\vartheta(\omega)$ by a constrained minimization, it is elementary 'geometry' in $\mathbb{C}$ to obtain

(3.12) $\vartheta_\nu(\omega) := \begin{cases} 1/\alpha & \text{when } 1/|\alpha| \leq s \\ s\bar{\alpha}/|\alpha| & \text{when } s \leq 1/|\alpha|. \end{cases}$

pointwise on Ω. This then gives $\nu \sup_\Omega\{|\vartheta|\} \leq \nu s$ and, for each $\omega \in \Omega$, gives $|1 - \vartheta\alpha| \leq (1 - s|\alpha|)_+$ where $(\cdot)_+$ here indicates the 'positive part'. Note that we need only consider $\omega \in \Omega$ for which $|\alpha\beta| = t \in \Sigma^+$; then, using (3.2), one has

(3.13) $|1 - \vartheta\alpha||\beta| \leq (|\beta| - s|\alpha\beta|)_+ \leq (\psi(t) - st)_+$

so the right hand side of (3.10) is bounded by $\sup_{t\in\Sigma^+}\{(\psi(t) - st)_+\}$; of course we have the bound for arbitrary choice of s. We observe that in taking the $\sup_t = \sup_\omega$ we also need never consider ω for which $1/|\alpha(\omega)| < s$, making $\psi(t) - st$ negative; thus the 'sup' remains the same if one omits taking the positive part. This expression depends lower semicontinuously on s and, on adding the coercive term νs, one need only consider s in a compact set so one does, indeed, have an attained minimum in the definition of $\hat{\varepsilon}_+(\nu)$ in (3.6); let $s_* = s_*(\nu)$ be such a minimizer. This shows that $\hat{\varepsilon}_+(\nu)$ is, indeed, a uniform bound realized by the linear filter given by (3.8) with (3.12), using $s = s_*$.

Using Lemma 1.1, we have now established the existence of $\varepsilon_\#$ and the upper bound in (3.5). We now obtain the lower bound: $\varepsilon_\# \geq \hat{\varepsilon}_-$. To this end, for any $\delta > 0$ we consider any $t_* \in \Sigma^+$ with $t_* \leq \nu$. For (1.6) with $\varphi_1 = 0$ and $\varepsilon_0 = \bar{\nu} := t_*$ we now consider

$$z(\cdot) := \frac{t_*}{t_* + \delta} \frac{\chi(\cdot)}{\sqrt{\operatorname{meas} S}}$$

where $S = S_{t_*;\delta}$ is as in (3.4) and χ is its characteristic function. Clearly, $\|z\| < 1$ and, since

$$\|\mathbf{AB}z\| = \|\alpha\beta z\| \leq \sup_S \{|\alpha\beta|\} \|z\|$$

and (3.4) gives $|\alpha\beta| < t_* + \delta$ on S (i.e., where $z \neq 0$), we have $\|\mathbf{AB}z\| \leq \bar{\nu}$ whence $x(\cdot) := \beta z$ is a solution. Since (3.4) also gives $|\beta| > [\psi(t_*) - \delta]$ where $z \neq 0$, we have $\|z\| \leq \|x\|/[\psi(t_*) - \delta]$. Now we have (1.3) for *both* the solutions x and $-x$ so we must have $2\varepsilon \geq \|x - (-x)\|$ whence

$$\varepsilon_\#(t_*) \geq \frac{t_*}{t_* + \delta}[\psi(t_*) - \delta].$$

Combining this with our choice of t_* and the obvious fact that $\varepsilon_\#(\bar{\nu}) \leq \varepsilon_\#(\nu)$ (since it is clear from (1.1) that $\varepsilon_\#(\cdot)$ is a nondecreasing function), we have the desired lower bound in the limit as $\delta \searrow 0$.

Finally, we consider the case of $\Sigma^+ = [0, \|\mathbf{AB}\|]$ with ψ increasing and concave. For increasing ψ defined on an interval, the definition of $\hat{\varepsilon}_-$ immediately gives $\hat{\varepsilon}_-(\nu) = \psi(\nu)$. Rewriting the argument of the min sup in defining $\hat{\varepsilon}_+$ as $[\psi(t) + s(\nu - t)]$, we recognize this definition as giving

$$\hat{\varepsilon}(\nu) := -[-\psi]^{**}(\nu)$$

where the $[\cdot]^*$ indicates Legendre-Fenchel conjugation. When ψ is concave so $-\psi$ is convex we have duality: $[-\psi]^{**} = -\psi$ (cf., e.g., [8]) so $\hat{\varepsilon} = \psi$ then. From (3.5) we now have

$$\psi(\nu) = \hat{\varepsilon}_-(\nu) \leq \varepsilon_\#(\nu) \leq \hat{\varepsilon}_+(\nu) = \psi(\nu)$$

which gives (3.7). With this, it is not difficult to see that the minimum occurs with $s = s_* := \psi'(\nu)$ when ψ is diffeentiable at ν and one has a comparable geometric characterization (by 'support') otherwise; this determines the $s = s_*$ to be used in (3.12).

Since the filter $\mathbf{F}_\nu$ defined above was shown to realize $\hat{\epsilon}_+(\nu)$ which is here the *minimal* uniform bound, this construction gives an optimal filter in this setting. ∎

4.   FURTHER REMARKS

REMARK 4.1:   We now assert that Theorem 3.1 gives $\epsilon_\#(\nu) \to 0$ as $\nu \to 0$ precisely when $t \to 0$ in Σ^+ implies $\psi(t) \to 0$. The necessity is immediate from the lower bound in (3.5). To see the sufficiency, we need only note that if we can obtain $\psi(t) < \bar{r}$ by requiring $t < \delta$ then, considering $s = \bar{r}/\delta$ for which $(\psi(t) - st)_+$ vanishes whenever $\psi(t) > \bar{r}$, we see that $\hat{\epsilon}_+(\nu) \le \nu\bar{r}/\delta + \bar{r}$ — if we can take this $\bar{r}$ arbitrarily small then $\hat{\epsilon}_+(\nu)$ can also be made arbitrarily small by, e.g., taking $\nu \le 1/\delta$.

By comparison with Theorem 2.3 we note that it is sufficient (but not necessary) for this that $\mathbf{B}$ be compact. When $\mathbf{B}$ is compact so is $\mathbf{AB}$ and we have a common sequence of eigenvectors $\{e_k\}$ with corresponding eigenvalues α_k and β_k, each converging to 0. Now $\rho(r) = \max\{|\beta_k| : |\alpha_k\beta_k| = r\}$ and, since there are only finitely many $|\beta_k| \ge \delta$ for any $\delta > 0$, we must eventually have only $|\beta_k| < \delta$ as $r \to 0$ eliminates consideration of those. ∎

REMARK 4.2:   It is interesting to note that instead of (3.10) one could have estimated by

(4.1)
$$\begin{aligned}\|x - \mathbf{F}\varphi_1\|^2 &\le \int_\Omega (|\beta||1 - \vartheta\alpha||z| + |\vartheta||\nu||\zeta|)^2 \\ &\le 2\sup_\Omega \{|\beta|^2|1 - \vartheta\alpha|^2 + \nu^2|\vartheta|^2\}\end{aligned}$$

and then modified the optimization procedure (3.11) to choose $\vartheta = \vartheta_\nu(\cdot)$ so as to minimize this last expression pointwise on Ω. It is elementary to minimize over $\vartheta \in \mathbb{C}$ and obtain $\vartheta = \lambda/\alpha$ with $0 \le \lambda := c^2/(1 + c^2) < 1$ where $c := |\alpha\beta|/\nu$. This gives

(4.2)
$$\vartheta_\nu := \frac{\bar{\alpha}|\beta|^2}{\nu^2 + |\alpha\beta|^2}$$

pointwise on Ω so

(4.3) $$|\beta|^2|1 - \vartheta\alpha|^2 + \nu^2|\vartheta|^2 = \frac{2\nu^2|\beta|^2}{\nu^2 + |\alpha\beta|^2}.$$

Using this and (3.2) in (4.1), we now obtain

(4.4) $$\varepsilon_\#(\nu) \leq \tilde{\varepsilon}_+(\nu) := \sup_{t \in \Sigma^+} \left\{ \frac{\sqrt{2}\nu\psi(t)}{\sqrt{\nu^2 + t^2}} \right\}.$$

This, of course, shows that $\tilde{\varepsilon}_+(\nu)$ also is a uniform bound which can be realized by the linear filter given by (3.8) with (4.2) rather than (3.12).

It is not immediately obvious how this result compares in general with the filter considered in Theorem 3.1. We *do* note that we necessarily have $\tilde{\varepsilon}_+(\nu) \geq \psi(\nu)$ since one can consider $t = \nu$ in the defining 'sup'. Thus, in the setting of (3.7) at least, this filter cannot be better than the filter of Theorem 3.1. For a special case, see Remark 4.3, below. ∎

REMARK 4.3:   The setting of (3.7) in Theorem 3.1 is particularly applicable to the interesting situation in which, for some $\lambda > 0$, one has

(4.5) $$|\mathbf{B}| = |\mathbf{A}|^\lambda, \qquad \text{i.e., } |\beta(\cdot)| = |\alpha(\cdot)|^\lambda \text{ ae on } \Omega.$$

Here, assuming for simplicity that $\Sigma^+ = \text{ess ran}(|\alpha\beta|)$ is an interval, one can carry out the computations by elementary calculus, without direct reference to (3.7). Observe that (4.5) gives, ae on Ω, that

$$|\beta(\cdot)| = |\alpha(\cdot)\beta(\cdot)|^\mu \qquad \text{with } \mu := \frac{\lambda}{1 + \lambda}$$

so we have $\psi(t) = t^\mu$ for $0 \leq t \leq \|\mathbf{AB}\|$. As $0 < \lambda$ gives $0 < \mu < 1$, we note that ψ is here (strictly) increasing and concave. Obviously $\hat{\varepsilon}_-(\nu) = \psi(\nu)$. Elementary calculus and a bit of algebraic manipulation shows that $(t^\mu - st)$ attains its maximum at $t = t_*(s) := (\mu/s)^{1/(1-\mu)}$, giving $qs^{-\lambda}$ with $q := (1 - \mu)\mu^\lambda$, and then that $\nu s + qs^{-\lambda}$ attains its minimum at $s = s_* = s_*(\nu) := (\lambda q/\nu)^{(1-\mu)}$, giving $\hat{\varepsilon}(\nu) = (\lambda + 1)qs_*^{-\lambda}$. With somewhat more manipulation, one simplifies these expressions to obtain

(4.6) $$s_*(\nu) = \mu\nu^{1-\mu}, \qquad \hat{\varepsilon}_+(\nu) = \nu^\mu.$$

In (3.12) we then use $s = s_*(\nu)$ from (4.6) to obtain the filter factor ϑ. Finally, we confirm that this has given $\hat{\varepsilon}_+(\nu) = \psi(\nu)$, just as asserted by (3.7) and also observe that it gives the 'log convexity' estimate

(4.7) $$\bar{\varepsilon}(\nu) = \varepsilon_0\varepsilon_\#(\nu) = \varepsilon_0^{1-\mu}\varepsilon_1^\mu$$

for the *original* problem (1.1).

One can similarly carry out the computations for the filter of Remark 4.2. Again elementary calculus suffices for computation of the maximization in (4.4) defining $\tilde{\varepsilon}_+$ and one easily sees that $\tilde{\varepsilon}_+(\nu) = C\nu^\mu$ with

$$C = C_\mu := \max\left\{\frac{\sqrt{2}t^{1-\mu}}{\sqrt{1+t^2}}\right\} = \sqrt{2\mu^\mu(1-\mu)^{1-\mu}}.$$

Note that $C_\mu \geq 1$ with equality only for $\mu = \frac{1}{2}$, corresponding to $\lambda = 1$. Thus, at least for this setting with $\lambda \neq 1$, the original filter is strictly better since it is optimal and we have $\hat{\varepsilon}_+ = \varepsilon_\# = \nu^\mu < C\nu^\mu = \tilde{\varepsilon}_+$. ■

REMARK 4.4:   Beside 'Problem 1', given by (1.1), one can consider another setting, 'Problem 2', involving a *parametrized family of such problems*. Thus, one might have a single space $\mathcal{X}$ but an *evolution family* of continuous linear operators[8] $\mathbf{S}(t,s) : \mathcal{X} \to \mathcal{X}$ with

$$(4.8) \qquad \begin{cases} \mathbf{S}(t,t) = \mathbf{I} \\ \mathbf{S}(t,s) \circ \mathbf{S}(s,r) = \mathbf{S}(t,r) & \text{for } t_0 \leq r \leq s \leq t \leq t_1. \end{cases}$$

One now has a parametrized family of problems like (1.1): seek a function $x(t) := \mathbf{S}(t,t_0)x_0$ for $t_0 < t < t_1$ such that

$$(4.9) \qquad \begin{aligned} &.1) \quad &&\|x(t_0) - \varphi_0\| \leq \varepsilon_0 \\ &.2) \quad &&\|x(t_1) - \varphi_1\| \leq \varepsilon_1 \end{aligned}$$

For each fixed t this has the same form as (1.1) with $\mathbf{A} = \mathbf{S}(t,t_1)$ and $\mathbf{B} = \mathbf{S}(t_0,t)$ and we note consistency (for all t with $t_0 < t < t_1$) provided there is some x_0 such that $\|x_0 - \varphi_0\| \leq \varepsilon_0$ and $\|\mathbf{S}(t_1,t_0)x_0 - \varphi_1\| \leq \varepsilon_1$. We now have the solution set $\mathcal{U}(t) := \mathcal{U}_0(t) \cap \mathcal{U}_1(t)$ where

$$(4.10) \quad \begin{aligned} \mathcal{U}_0(t) &:= \{x \in \mathcal{X} : x = \mathbf{S}(t,t_0)x_0 \text{ with } \|x_0 - \varphi_0\| \leq \varepsilon_0\}, \\ \mathcal{U}_1(t) &:= \{x \in \mathcal{X} : \|\mathbf{S}(t_1,t)x - \varphi_1\| \leq \varepsilon_1\}. \end{aligned}$$

In the linear case we obviously have the same reduction as in Lemma 1.1 to replace consideration of (4.9) in general by consideration after replacing $[\varphi_0, \varepsilon_0]$ by $[0, 1]$ while also replacing $[\varphi_1, \varepsilon_1]$ by $[\tilde{\varphi}, \nu]$ with $\nu := \varepsilon_1/\varepsilon_0$ and with $\tilde{\varphi} := (\varphi_1 - \mathbf{S}(t_1, t_0)\varphi_0)/\varepsilon_0$. We then seek a filter — more precisely, a family of filters parametrized by $t \in [t_0, t_1]$ — as a map

$$\mathbf{F} = \mathbf{F}_\nu(t) : \mathcal{H} \to \mathcal{H} : \tilde{\varphi} \mapsto x_*(t)$$

for which we obtain a uniform bound

$$(4.11) \qquad \|\mathbf{S}(t,t_0)x_0 - \mathbf{F}_\nu(t)\tilde{\varphi}\| \leq \varepsilon(\nu, t)$$

$$\text{for } x_0 \in \mathcal{B} \text{ with } \|\mathbf{S}(t_1,t_0)x_0 - \tilde{\varphi}\| \leq \nu.$$

---

[8]One could also, more generally, be treating nonlinear maps $\mathbf{S}(t,s) : \mathcal{X}_s \to \mathcal{X}_t$ with a scale of spaces $\{\mathcal{X}_t : t_0 \leq t \leq t_1\}$.

We call such a filter $\mathbf{F}_\nu(t)$ an **optimal filter** if $\varepsilon(\nu,t) = \varepsilon_\#(\nu,t)$ is minimal (for $0 < \nu$ and $t_0 < t < t_1$).

For the present, we comment on this situation only in the context of a *semigroup* $[\mathbf{S}(t,s) = \mathbf{S}(t-s)]$ of normal linear operators acting on a Hilbert space $\mathcal{H}$. Note that this fits the case of (4.5) since for each t ($t_0 < t < t_1$) one has $\mathbf{A} = \mathbf{S}(t_1 - t)$, $\mathbf{B} = \mathbf{S}(t - t_0)$ and, using semigroup properties, one has $\mathbf{B} = \mathbf{A}^\lambda$ with

$$\lambda = \frac{t - t_0}{t_1 - t} \qquad \text{giving } \mu := \lambda/(1+\lambda) = \frac{t - t_0}{t_1 - t_0}.$$

For notational convenience we now assume t is normalized to give $t_0 = 0$ and $t_1 = 1$ so $\mu = t$. It is also convenient to use an exponential formulation so we write our spectral decomposition in the form:

$$\beta(\cdot) = e^{-t\gamma(\cdot)}, \quad \alpha(\cdot) = e^{-(1-t)\gamma(\cdot)} \quad \text{with } \gamma(\cdot) = g(\cdot) + ih(\cdot)$$

where the function $\gamma(\cdot)$ represents the infinitesimal generator of the semigroup. Note that Σ^+ is now $\exp[-\operatorname{ess\,ran}(g)]$ and that our filter has been shown optimal when this is an interval. Using (4.6) in the form

$$(4.12) \qquad s_* = s_*(\nu,t) = e^{(1-t)\bar{r}} \quad \text{with } \bar{r} = \bar{r}_\nu(t) := \log\nu + \frac{\log t}{1-t},$$

the filter $\mathbf{F}_\nu(t)$ here becomes multiplication by ϑ where

$$
\begin{aligned}
\vartheta \;&=\; \vartheta_\nu(t,\cdot) := e^\rho \qquad \text{with} \\
(4.13) \qquad \rho \;&=\; \rho_\nu(t,\omega) \\
&:=\; i(1-t)h(\omega) +
\begin{cases}
(1-t)g(\omega) & \text{when } g(\omega) \le \bar{r}_\nu(t), \\
(1-t)\bar{r} & \text{otherwise.}
\end{cases}
\end{aligned}
$$

As noted, this is an optimal filter, realizing the minimal uniform bound ν^t (also, see (4.7)). ∎

REMARK 4.5:   The 'sideways heat equation' problem treated in [3] fits the setting of Remark 4.4 and it is interesting to note that we have obtained the identical result here as there although the derivation was rather different. One may also compare our derivation of (4.7) with [7]. Here, the derivation was through the relation to the 'stationary' problem (for one t at a time, essentially independently). In [3], on the other hand, the emphasis was 'dynamic' — marching with respect to the parameter against the smoothing effect of the semigroup. This meant, essentially, that the same approach as in (3.11) was applied 'differentially' in the context of a problem with 'moving information'. In our present notation we may formulate that as: at each t one was 'given' $[v(t + dt), \nu(t + dt)]$ and considered the problem of finding the pair $[v(t), \nu(t)]$ so as to approximate $x(t)$ satisfying

(4.14)   .1) $x(t) = \mathbf{S}(t)\tilde{z}$ with $\tilde{z} \in \mathcal{B} := \{z \in \mathcal{Z} : \|z\| \leq 1\}$

    .2) $\|\mathbf{S}(dt)x(t) - v(t + dt)\| \leq \nu(t + dt)$

with an error bound $\nu(t)$. This then led to a differential inequality for $\nu(\cdot)$ (considered for t progressing 'leftward' from $t = 1$, using the given values $[v(1), \nu(1)] = [\tilde{\varphi}, \nu]$, towards $t \searrow 0$) in which $\bar{r}(t)$ is viewed as a 'control', optimally chosen to make $\nu(\cdot)$ grow as slowly as possible. Thus, $\bar{r}$ becomes a function of $\nu(t)$, much as in (4.12) but modified in view of the differential re-scaling of the parameter interval. It turns out that the differential equations can be solved explicitly, leading to explicit formulas for $[v(t), \nu(t)]$ which coincide with those obtained in Remark 4.4. The backwards heat equation can be treated similarly [4].

While the present treatment may appear conceptually simpler for the derivation in this case, it does seem that the differential approach employed in [3] may be effective in the generation of useful (suboptimal) algorithms for cases involving evolutionary families not of semigroup form — so one may not have the commutativity implicit in the spectral resolution used here (or even linearity) but might have these to a good enough approximation differentially, i.e., by linearization, ... ∎

References

[1] J.R. Cannon, *A priori estimate for continuation of the solution of the heat equation in the space variable*, Ann. Mat. Pura Appl. **65** (1964), pp. 377–388.

[2] L. Eldén, *Hyperbolic approximations for a Cauchy problem for the heat equation*, Inverse Problems **4** (1988), pp. 59–70.

[3] L. Eldén and T.I. Seidman, *An 'Optimal Filtering' method for the sideways heat equation*, to appear.

[4] R. Ewing and T.I. Seidman, *'Optimal Filtering' for the backwards heat equation*, to appear.

[5] F. John, *Continuous dependence on data for solutions with a prescribed bound*, Comm. Pure Appl. Math. **13**, (1960).

[6] V. Klee, in Math. Rev. **13** (1952), p. 661.

[7] H.A. Levine, *Continuous data dependence, regularization, and a three lines theorem for the heat equation with data in a space like direction*, Ann. Mat. Pura Appl. (IV) **CXXXIV** (1983), pp. 267–286.

[8] R.T. Rockafellar, *Convex analysis*, Princeton Uni. Press, Princeton, 1970.

[9] W. Rudin, *Functional Analysis*, McGraw–Hill, New York, 1973.

[10] I.E. Segal and R.A. Kunze, *Integrals and Operators*, McGraw–Hill, New York, 1968.

[11] T.I. Seidman, *An inverse eigenvalue problem with rotational symmetry*, Inverse Problems 4 (1988), pp. 1093–1115.

[12] T.I. Seidman, *Some 'complexity' issues for ill-posed problems*, in Proc. AMS–SIAM Conference on Ill-Posed and Inverse Problems (D. Colton, R. Ewing, W. Rundell, eds.), SIAM, 1990.

[13] R. Temam, *Navier-Stokes Equations, Theory and Numerical Analysis*, North-Holland, Amsterdam, 1984.

CHAPTER 9

Parallel Pseudospectral Methods for the Solution of the Wave Equation*

R. A. Renaut†
M. L. Woo†

Abstract. Pseudospectral methods have become popular in recent years for the solution of many problems in aerodynamics and fluid mechanics because of their high order accuracy and good resolution. In this paper we discuss the numerical solution of the acoustic and elastic wave equations on a hypercube architecture using a Fourier pseudospectral method. The efficiency depends on the choice of the parallel fast Fourier transform. We present an algorithm for the FFT that uses d communications on a cube of dimension d. Data are not assigned to the nodes in a natural ordering as in the algorithms proposed by Swarztrauber [7] and Chamberlain [1]. In this way we save one node to node communication compared to [7] and use shorter message lengths compared to [1]. Numerical results for the solution of the first order hyperbolic equations on the Intel iPSC 1 hypercube demonstrate that our algorithm is both load balanced and efficient. Work on the iPSC2 is in progress.

1. Introduction. Pseudospectral methods offer an alternative to high accuracy finite difference methods for the solution of partial differential equations. They can be used to improve accuracy and high frequency resolution while at the same time the numerical size of the problem is reduced. Kosloff and Baysall [5] introduced the use of the Fourier pseudospectral method for the solution of the acoustic wave equation and Fornberg [4] looked at the two dimensional elastic wave equation. In either case spatial derivatives are found using one-dimensional trigonometric approximations separately in both x- and y-directions. For example, we briefly describe the method of [5] for the two-dimensional acoustic wave equation:

---

* The work of the first author was supported by National Science Foundation grant ASC-8812147 and an American Chemical Society Petroleum Research Fund grant 20681-G2.

†Department of Mathematics, Arizona State University, Tempe, AZ 85287–1804

$$\frac{\partial}{\partial x}\left(\frac{1}{\rho}\frac{\partial P}{\partial x}\right) + \frac{\partial}{\partial y}\left(\frac{1}{\rho}\frac{\partial P}{\partial y}\right) = \frac{1}{c^2\rho}\frac{\partial^2 P}{\partial t^2} \qquad (1-1)$$

In this equation $P(x, y, t)$ represents the pressure, $\rho(x, y)$ the density and $c(x, y)$ the wave velocity.

We assume that $P(x, y, t)$ is required on a periodic domain in x and y. A pseudospectral Fourier approximation may then be used for the spatial derivatives in (1-1). Integration in time is carried out using a standard ordinary differential equation solver such as leapfrog, with timesteps taken small enough not to effect the stability of the numerical solution. Thus (1-1) converts to the system of ordinary differential equations

$$\frac{d^2 P(i, j, t)}{dt^2} = \mathcal{L}P(i, j, t) \qquad (1-2)$$

where $\mathcal{L}$ represents the operator describing the spatial terms and $P(i, j, t) \approx P(i\Delta x, j\Delta y, t)$. The right-hand side of (1-2) is calculated in two separate passes, one for the term containing the x-derivatives and one for the term with y-derivatives.

The operator $\mathcal{L}$ can be obtained in various ways. Here, suppose that the solution is required on the domain $[x, y] \in [0, 1] \times [0, 1]$. The x-points are distributed on $[0, 1]$ according to $x_j = j/2N$, $j = 0, 1, \ldots, 2N - 1$. Then, supposing that the function $P(x)$ defined by the values of $P(x, y, t)$ at the points x_j, is a smooth differentiable function we can approximate it by a trigonometric polynomial interpolating at the x_j of the form

$$\tilde{P}_N(P(x)) = \sum_{j=0}^{2N-1} P(x_j)g_j(x). \qquad (1-3)$$

Here $g_j(x_k) = \delta_{jk}$ and $g_j(x)$ is a trigonometric polynomial of degree N:

$$g_j(x) = \frac{1}{2N}\sum_{l=-N}^{N}\frac{1}{c_l}e^{il(x-x_j)}, \qquad (1-4)$$

$$c_l = 1(|l| \neq N), \quad c_N = c_{-N} = 2.$$

Equivalently,

$$\tilde{P}_N(P(x)) = \sum_{l=-N}^{N} a_l e^{ilx},$$

$$a_l = \frac{1}{2Nc_l}\sum_{j=0}^{2N-1} P(x_j)e^{-ilx_j}. \qquad (1-5)$$

The derivative of $P(x)$ is found from

$$\frac{d}{dx}(\tilde{P}_N(x_j)) = \sum_{l=-N}^{N} ila_l e^{ilx_j}. \qquad (1-6)$$

In (1-1) two x-derivatives are required. Initially $\frac{\partial P}{\partial x}$ is calculated along each x-grid line by performing a spatial fast Fourier transform (FFT) on $P(x)$ to obtain the coefficients a_l given by (1-5). These coefficients are then complex multiplied by the wave number il and the derivatives evaluated with the inverse FFT (1-6). At the next stage the derivative $\frac{\partial P}{\partial x}$ is vector multiplied by $\frac{1}{\rho(x)}$ and again a forward and inverse FFT are used to obtain $\frac{\partial}{\partial x}(\frac{1}{\rho}\frac{\partial P}{\partial x})$. The same procedure is used to obtain the y-derivatives and hence a total of four sweeps, each comprising an FFT and its inverse, are required.

Apparently, the calculation of the FFT and its inverse is the major expense in the implementation of the Fourier pseudospectral method. For different formulations of the problem, e.g., Fornberg [4], different numbers of transforms are required, but still comprise the major expense. Thus, in parallel an efficient one-dimensional FFT and its inverse is required. There are, however, a number of options available for the solution of (1-1) in parallel on a hypercube. The solution approach depends on the mapping of the physical domain to the hypercube architecture. Chu [2] discusses the options for two-dimensional FFTs on a hypercube. Although the solution of (1-1) only needs one-dimensional FFT's it is a two-dimensional problem and, as such, Chu's discussion is appropriate. If the physical domain is mapped stripwise to the hypercube so that x-grid lines are contained entirely within one processor but y-grid lines are distributed we can use either a transpose-split (TS) or a local distributed (LD) algorithm. In the former case the x direction FFT's are performed locally and then the whole domain is transposed and the y derivatives are calculated locally. Thus, in this case, no parallel FFT algorithm is needed but an efficient transpose algorithm is a prerequisite. For the LD algorithm the x-derivatives are found locally, as above, but the y-derivatives are calculated from a distributed FFT algorithm. Alternatively the physical domain can be mapped to the hypercube with subdomains mapped to individual nodes so that both x and y directions are distributed across the cube. This gives the block FFT for which distributed FFT's in both directions are necessary. We dismiss the TS algorithm for implementation on the iPSC1 because of the transpose operation which is known to be inefficient on the iPSC1 unless programmed carefully and concentrate on the distributed FFT which is required in the other cases. We note, however, that this conclusion is not valid on the second generation hypercube. With the iPSC1 it is crucial that communications are restricted as much as possible to be between nearest-neighbor nodes. It is claimed that the communications hardware and software on the iPSC1 make these considerations irrelevant. Thus the TS algorithm may be adequate on the iPSC2. We are currently implementing our FFT on the iPSC2 and will investigate the relative efficiencies of the TS and LD algorithms.

In section 2 we describe the FFT and how it can be implemented in parallel. Some choices are considered and we argue why our algorithm is the most efficient. In section 3 numerical results are presented which demonstrate the algorithm's efficiency and good load balancing. Tests are based on the solution of the one-dimensional hyperbolic equation with periodic initial condition. Work on more general problems is in progress.

2. The Parallel FFT. Cooley and Tukey [2] first proposed the Fast Fourier transform for the computation of the discrete Fourier transform (DFT). For a series of length N they showed that the DFT could be calculated in order $N \log N$ operations. In general N is any number but for many applications it is taken to be a power of two. We suppose that N is a power of two as this is natural for the power of two topology of the hypercube.

Let there be $N = 2^r$ data points denoted by $x_0, x_1, \ldots, x_{N-1}$. The DFT of this sequence is given by

$$X_k = \sum_{n=0}^{N-1} x_n e^{\frac{-ink2\pi}{N}}. \qquad (2-1)$$

Since $N = 2 \cdot 2^{r-1}$ we can use the standard index maps

$$n = i + 2j, \quad k = l + 2^{r-1}m \qquad (2-2)$$

and define the two-dimensional arrays

$$x(i,j) = x_n, \qquad i = 0, 1, j = 0, 1, \dots, 2^{r-1}$$
$$X(l,m) = x_k, \qquad l = 0, \dots, 2^{r-1} - 1, m = 0, 1. \tag{2-3}$$

Substituting (2-2) and (2-3) into (2-1) we obtain

$$X(l,m) = \sum_{i=0}^{1} \omega_2^{im} \omega_N^{il} \sum_{j=0}^{2^{r-1}-1} x(i,j) \omega_{2^{r-1}}^{jl} \tag{2-4}$$

where $\omega_p = e^{\frac{-i2\pi}{p}}$. Thus $X(l,m)$ and hence X_k can be computed as two multiple transforms using

$$X^1(i,l) = \omega_N^{il} \sum_{j=0}^{2^{r-1}-1} x(i,j) \omega_{2^{r-1}}^{jl}, \quad i = 0, 1 \tag{2-5}$$

and

$$X^2(m,l) = \sum_{i=0}^{1} X^1(i,l) \omega_2^{im}, \quad m = 0, 1, \dots, 2^{r-1} \tag{2-6}$$

Therefore $X_k = X(l,m) = X^2(m,l)$ is obtained by digit reversal. Since 2^{r-1} now factors by 2 (2-5) can be calculated by reapplying the above technique. In general, for $N = 2^r$, the FFT has the following properties:

(i) r butterflies are required to compute the summation, where a butterfly is the process from X^i to X^{i+1}, $i = 0, 1, \dots, r$.

(ii) after r butterflies the order of the data changes and the data id is exactly the bit reverse of the original data id.

(iii) the data can be reordered with just one switch. This may be done before or after the butterflies.

Observe, however, that when both an FFT and its inverse are performed on the data, step (iii) above is unnecessary. The inverse FFT then uses a different ordering to the FFT, the ordering chosen so that it is the order of the scrambled data on exit from the FFT. On a parallel machine step (iii) is performed using an across node communication. Therefore two node to node communications are saved per FFT and inverse pair. Swarztrauber [7] states that such an unordered transform requires $d+1$ communications on a cube of dimension d. In that paper he assigns the data to the nodes in a natural ordering, i.e., the data is divided into blocks and the first block is assigned to node zero and so on. At each global butterfly each node communicates half of its entries to the appropriate neighboring node. Chamberlain [1], on the other hand, also uses the natural ordering to assign the data to the cube but at each global butterfly each node communicates all of its entries to the appropriate neighboring node. Then only d communications are required on a cube of dimension d. Clearly, Chamberlain's algorithm has the advantage of fewer parallel transmissions but the disadvantage that the message sizes are larger. We have developed an algorithm with the advantages of both of the algorithms just described.

We have found that by allocating the data to the nodes in an "optimal" fashion the communication is minimized. Since the inverse FFT acts on data in a different order to the FFT two algorithms are required. For the FFT the data is assigned according to the relationship dataid $=$ node id $+ j \cdot 2^d$, $j = 0, 1, \dots, 2^k - 1$, where $k = r - d$. For example with 4 nodes and 16 data the data are allocated as follows:

node (0)	0	4	8	12
node (1)	1	5	9	13
node (2)	2	6	10	14
node (3)	3	7	11	15.

Then the first k butterflies are performed locally and the remaining d are global. The input to the inverse FFT is the bit reverse of the unordered output of the FFT. Thus data are assigned according to

$$\text{data id} = (\text{bit reverse (node id)} + j \times 2^r) \times 2^{k-1} + l,$$
$$l = 0, 1, \dots, 2^{k-1} - 1, \quad j = 0, 1.$$

Here the inverse FFT does 1 local butterfly followed by d global butterflies and $k-1$ local butterflies. From Figures 2.1 and 2.2 we see that only half of the data points in each node are transformed in each global butterfly and that communication is between neighboring nodes. To save computation we precompute the values of the weights ω^j required for the FFT and its inverse. This is necessary in the solution of partial differential equations because we integrate over many timesteps and repetitive computation would be costly.

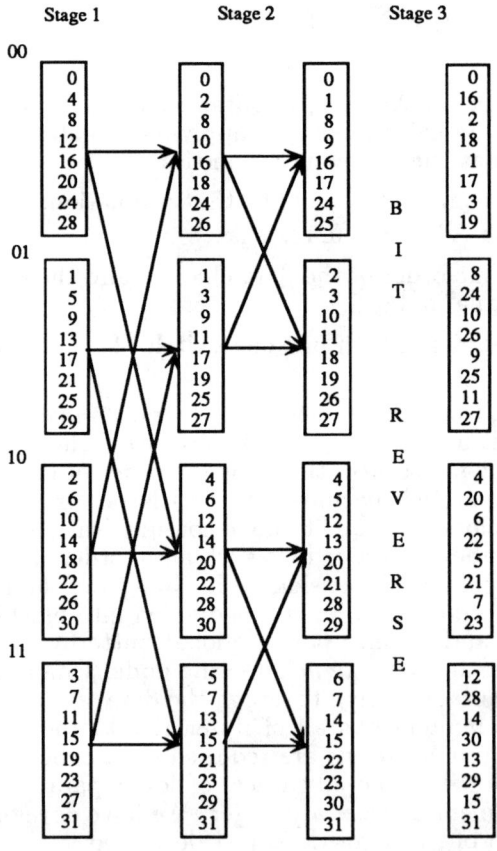

FIG. 2.1 *The global butterflies for the FFT in the case of 4 nodes and 32 data points*

both the fastest and slowest nodes was recorded in order to detect load imbalance. Measurements are in milliseconds to the nearest 5 milliseconds. Recordings were made over both 2 and 15 timesteps and it was found that the average time per timestep was very close. The averages from results over 2 timesteps are presented in Table 3.1. Note that these are then the average times for an FFT pair. Compared with Table 1 in [1] we see that our algorithm is much better load balanced. Also, the last column in that table gives results for one FFT with powers of ω precalculated for a transform of length 4096. Supposing that an algorithm would require twice the time for 4096 nodes as for 2048 we also see that our algorithm is faster (note that we also divide by two since our results are for the FFT pair). In fact we should expect to multiply by a factor just greater than two as we would not expect perfect efficiency. In either case our results are better. For comparison we have tabulated these figures in Table 3.1. The comparison is valid as Chamberlains algorithm was also programmed on the Intel iPSC1.

Number of nodes	Number of data points									
	16	32	64	128	256	512	1024	2048	4096 predicted	4096 Chamberlain
1	38	98	240	580	1395	3298	7708	17843	35686	38080
2	23 / 25	50 / 53	115 / 115	270 / 273	638 / 640	1498 / 1510	3505 / 3523	8115 / 8163	16230 / 16326	18336 / 20160
4	20 / 20	33 / 33	63 / 65	135 / 138	305 / 308	698 / 710	1623 / 1645	3748 / 3793	7496 / 7586	8832 / 10592
8	20 / 20	40 / 43	40 / 43	95 / 103	155 / 170	340 / 365	763 / 815	1765 / 1870	3530 / 3740	4352 / 5600
16	25 / 28	70 / 73	73 / 78	100 / 108	115 / 130	175 / 205	375 / 433	828 / 943	1656 / 1886	2112 / 2944
32		30 / 33	98 / 103	105 / 115	123 / 125	133 / 163	215 / 273	410 / 523	820 / 1046	1024 / 1606

TABLE 3.1

Average times (fastest/slowest) per time step.

In Figures 3.1 and 3.2 we present the speed up and efficiency,

$$\text{speed up} = \frac{\text{time for 1 node}}{\text{time for } N \text{ nodes}}$$

and

$$\text{efficiency} = \frac{\text{cost of 1 node}}{\text{cost of } N \text{ nodes} \times N}.$$

Stage 1   Stage 2   Stage 3

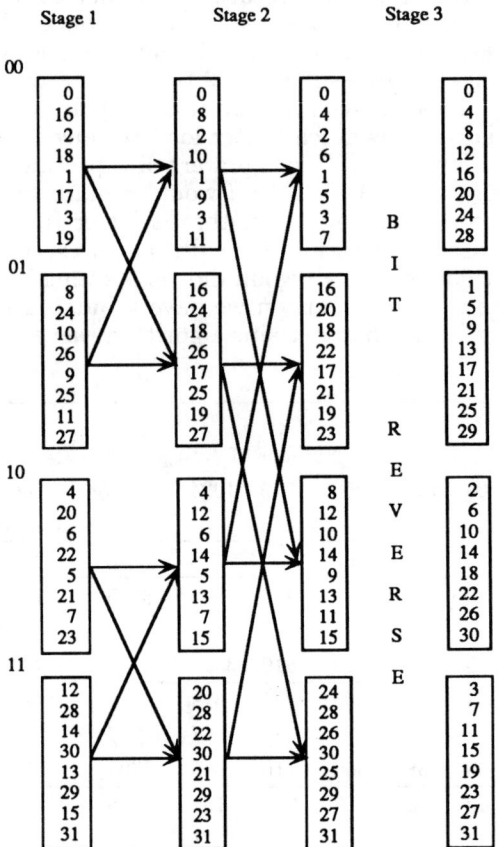

FIG. 2.2 *The global butterflies for the inverse FFT in the*

case of 4 nodes and 32 data points

The only disadvantage that we see with our algorithm is that the data are not ordered in a natural way. This means that results must be carefully organized on output from the program. Note, however, that the natural ordering may also not be appropriate in many cases. Chamberlain [1] also compared his FFT to an FFT based on a Gray code. Communications in the FFT are then between nodes a distance of two apart but neighboring parts of the domain are in neighboring processors. This can be beneficial in many physical problems where information between neighboring domains must be exchanged. We shall investigate whether the ordering in our algorithm causes problems in further work.

3. Results. We tested our algorithm for the solution of the hyperbolic equation

$$u_t = u_x \qquad t > 0, \quad 0 < x < 2\pi \qquad (3-1)$$

with periodic initial condition $u(x,0) = \sin x$, $0 < x < 2\pi$. In this case only one forward and one inverse FFT are required per timestep. Leapfrog was used to integrate in time. The program was run on the Intel iPSC1 at Argonne which has 32 nodes. We tested data structures of dimension 2^4 to 2^{11}. In each case the time for

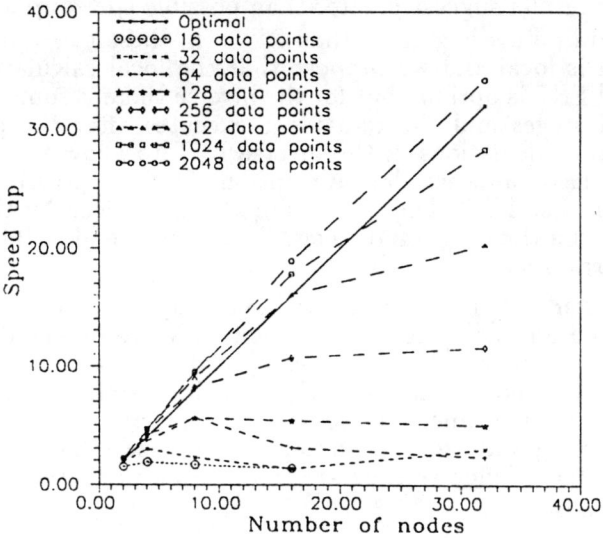

FIG. 3.1 *Speed up (1 node/N node) over 2 time steps*

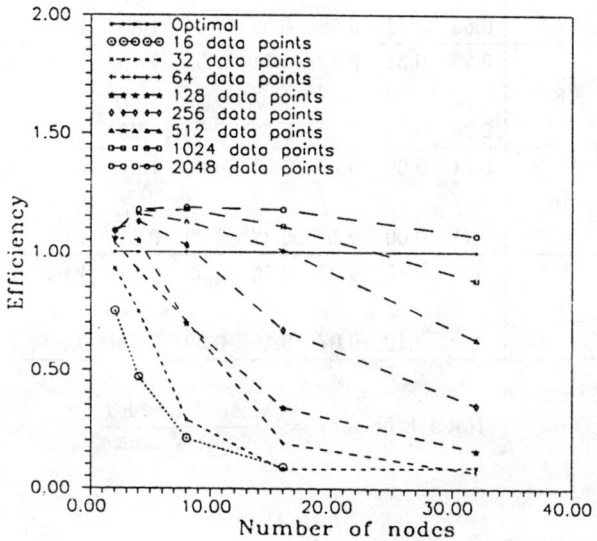

FIG. 3.2 *Efficiency (1 node/N node) over 2 time steps*

Theoretically optimal speed up is N and optimal efficiency 1. We observe, however, that our results suggest greater than possible speed up and efficiency when the number of data points is greater than $2^{d \times 5}$. In these cases proportionally more of the calculation is local and we propose that the local calculation is suboptimal. Actually the local FFT is optimal but for the inverse there is some part that executes before the global stages and the remainder executes after the global part. This requires some extra calculations in the inverse FFT to allow the algorithm to work for any cube dimension and any data dimension. It is not possible to overcome this problem for the inverse FFT. The results are also influenced by the very large data arrays required to run the program on a small number of nodes. The Intel iPSC deals with large data arrays less efficiently.

In Tables 3.2 and 3.3 the efficiency and speed up are calculated with respect to 2 nodes. Again the unusual speed-up and efficiency are observed. In Table 3.4 we present the size of the message in each case. It is known that the communication time on the Intel iPSC1 does not increase linearly with time, McBryan and van de Velde [6]. In fact the communication time increases discontinuously as message length increases by 1024 bytes. Messages are sent in packages up to size 1024 bytes. Our results are influenced by this inefficiency as they are slower than predicted for messages of size greater than 512 bytes.

Number of nodes	Number of data points							
	16	32	64	128	256	512	1024	2048
4	0.56 / 0.63	0.77 / 0.81	0.92 / 0.88	1.00 / 0.99	1.05 / 1.04	1.07 / 1.06	1.08 / 1.07	1.08 / 1.08
8	0.28 / 0.28	0.31 / 0.31	0.72 / 0.68	0.71 / 0.66	1.03 / 0.94	1.10 / 1.03	1.15 / 1.08	1.15 / 1.09
16	0.11 / 0.11	0.09 / 0.09	0.20 / 0.19	0.34 / 0.32	0.69 / 0.62	1.07 / 0.92	1.17 / 1.02	1.23 / 1.08
32		0.10 / 0.10	0.07 / 0.07	0.16 / 0.15	0.33 / 0.32	0.71 / 0.58	1.02 / 0.81	1.24 / 0.98

Note: Efficiency $= \dfrac{\text{Cost of 2 nodes}}{\text{Cost of } N \text{ nodes}}$

TABLE 3.2
Efficiency (fastest/slowest) over 2 time steps

Number of nodes	Number of data points							
	16	32	64	128	256	512	1024	2048
4	1.13/1.13	1.54/1.62	1.84/1.77	2.00/1.98	2.09/2.08	2.15/2.13	2.16/2.14	2.17/2.15
8	1.13/1.11	1.25/1.24	2.88/2.71	2.84/2.66	4.11/3.77	4.40/4.14	4.60/4.32	4.60/4.37
16	0.90/0.91	0.71/0.72	1.59/1.48	2.7/2.54	5.54/4.92	8.56/7.37	9.35/8.15	9.81/8.66
32		1.67/1.62	1.18/1.12	2.57/2.37	5.20/5.12	11.30/9.29	16.30/12.93	19.79/15.62

Note: Speedup $= \dfrac{\text{Time for 2 nodes}}{\text{Time for } N \text{ nodes}}$

TABLE 3.3

Speed up (fastest/slowest) over 2 time steps

Number of nodes	Number of data points							
	16	32	64	128	256	512	1024	2048
2	32	64	128	256	512	1024	2048	4096
4	16	32	64	128	256	512	1024	2048
8	8	16	32	64	128	256	512	1024
16	8	8	16	32	64	128	256	512
32		8	8	16	32	64	128	256

TABLE 3.4

*Length (bytes) of the message transmitted
per communication*

We conclude that our program is load balanced and efficient. It compares positively with the algorithm presented by Chamberlain [1]. Swarztrauber did not program his algorithm and so we have no comparison but theoretically our algorithm is superior. Thus the algorithm presented here provides a good tool for the efficient solution of partial differential equations using a parallel pseudospectral method.

References

[1] R. M. CHAMBERLAIN, *Gray Codes, Fast Fourier Transforms and Hypercubes*, Report no. CCS 86/1, Chr. Michelsen Inst., Fantoft, Norway, 1986.

[2] C. Y. CHU, *Comparison of Two-Dimensional FFT Methods on the Hypercube*, ACM, 7 (1988), pp. 1430–1437.

[3] J. W. COOLEY and J. W. TUKEY, *An Algorithm for the Machine Calculation of Complex Fourier Series*, Math. Comput. 19, 90 (1965), pp. 297–301.

[4] B. FORNBERG, *The Pseudospectral Method: Comparison with finite differences for the elastic wave equation*, Geophysics, 52, 4 (1987), pp. 483–501.

[5] D. D. KOSLOFF and E. BAYSALL, *Forward modeling by a Fourier method*, Geophysics, 47, 10 (1982), pp. 1402–1412.

[6] O. A. McBRYAN and E. F. van de VELDE, *Hypercube Algorithms and Implementations*, SIAM J. Sci. Stat. Comput., 8, 2 (1987), s227–s287.

[7] P. SWARZTRAUBER, *Multiprocessor FFT's*, Parallel Computing, 5 (1987), 197–210.